Students of human physiology are faced with many quantitative values such as ion concentrations, lung and fluid volumes, and blood pressures. In this book, using simple calculations, the reader is given opportunities to develop a feel for such values, to give them meaning and to derive fresh insights from them. These calculations are neither brain-teasers nor mere practical exercises, and require little more than easy arithmetic for their solution. Guidance is also given on the tricks of approximation and back-of-envelope arithmetic, and on coping with the varied systems of units that encumber physiology.

The main areas covered are energy metabolism, nerve and muscle, blood and the cardiovascular system, respiration, renal function, body fluids and acid–base balance. The book is intended as supplementary reading for students of physiology at any level, including graduates seeking a new slant on the subject.

PHYSIOLOGY BY NUMBERS

PHYSIOLOGY BY NUMBERS

An encouragement to quantitative thinking

R. F. BURTON

Senior Lecturer, Institute of Physiology, University of Glasgow

Published by the Press Syndicate of the University of Cambridge
The Pitt Building, Trumpington Street, Cambridge CB2 1RP
40 West 20th Street, New York, NY 10011-4211, USA
10 Stamford Road, Oakleigh, Melbourne 3166, Australia

First published 1994

Printed in Great Britain at the University Press, Cambridge

A catalogue record for this book is available from the British Library

Library of Congress cataloguing in publication data

Burton, R. F. (Richard F.)
Physiology by numbers : an encouragement to quantitative thinking
/ R. F. Burton.
p. cm.
Includes bibliographical references and index.
ISBN 0-521-42067-9 (hc).—ISBN 0-521-42138-1 (pb)
1. Physiology—Mathematics. I. Title.
QP33.6.M36B87 1994
612′.001′51—dc20 93-37134 CIP

ISBN 0 521 42067 9 hardback
ISBN 0 521 42138 1 paperback

KW

Contents

Preface

Let us therefore take it that in a man the amount of blood pushed forward in the individual heartbeats is half an ounce, or three drams, or one dram, this being hindered by valves from re-entering the heart. In half an hour the heart makes more than a thousand beats, indeed in some people and on occasion, two, three or four thousand. Now multiply the drams and you will see that in one half hour a thousand times three drams or two drams, or five hundred ounces, or else some such similar quantity of blood, is transfused through the heart into the arteries – always a greater quantity than is to be found in the whole of the body.

But indeed, if even the smallest amounts of blood pass through the lungs and heart, far more is distributed to the arteries and whole body than can possibly be supplied by the ingestion of food, or generally, unless it returns around a circuit.

William Harvey, *De Motu Cordis*, 1628 (from the Latin)

In more familiar terms, if the heart beats, say, 70 times a minute, ejecting 70 ml of blood into the aorta each time, then more fluid is put out in half an hour (147 l) than is either ingested in that time or contained in the whole of the body. Therefore, the blood must circulate. Thus may the simplest calculation bring understanding. I invite the reader to join me in putting two and two together likewise, hoping that my collection of simple calculations will also bring enlightenment.

Although my main aim is to share some insights into physiology obtained through calculation, I have written also for those many students who seem to rest just on the wrong side of an educational threshold – knowing calculators and calculus, but shy of arithmetic; drilled in accuracy and unable to approximate; unsure what to make of all those physiological concentrations, volumes and pressures that are as meaningless as telephone numbers until toyed with, combined, or re-expressed. As 'an encouragement to quantitative thinking' I also offer, for those ill at ease with arithmetic, guidance on how to cheat at it, cut corners, and be not too concerned for spurious accuracy. Harvey's calculations illustrate very well that a correct conclusion may be reached in spite of considerable inaccuracy. In his case it was the estimate of cardiac output that was wrong; it is now known to be about two-and-a-half ounces per beat. (There are eight drams to the ounce.)

Much of physiology requires precise computation, so I must not appear too much the champion of error and slapdash. There are, however, situations where even the roughest of calculations may suffice. Consider the generalization (see Section 2.8) that small mammals have higher metabolic rates per unit body weight than do large ones: taking the case of a hypothetical mouse with the metabolic rate of a steer, Max Kleiber (1961) calculated that to keep in heat balance in an environment at 3 °C, its surface covering, if like that of the steer, would need to be at least 20 cm thick! Arguments of this kind appear below. Be warned, however, that improbable answers are not always wrong, as exemplified by Rudolph Heidenhain's calculation of glomerular filtration rate in 1883 (see Section 5.5).

The book is based on an assortment of questions to be answered by calculation, together with some introductory and background information and comment on the answers. (The answers are given at the end of the book, together with notes and references.) Such a quantitative approach is more suited to some areas of physiology than to others and the coverage of the book naturally reflects this. The book is neither a general guide to basic physiology, nor a collection of brain-teasers or practice calculations. It rarely strays from shopkeeper's arithmetic and it is not a primer of mathematical physiology or of mathematics for physiologists. Rather, it is supplementary thinking for those who have done, or are still doing, at least an elementary course in physiology. I have learned much myself from the calculations and hope that other mature students may learn from them too.

Except where otherwise stated, the calculations refer to the human body. This is often taken as that of the physiologist's standard 70 kg adult man and many 'standard' textbook quantities are used here. This is partly to reinforce them in the reader's memory and build bridges from one to another, but such standard values are also a natural starting point for back-of-envelope calculations. Indeed, if there is any virtue to learning these quantities, it is surely helpful to exercise them and put them to use. Thus may one hope to bring life to numbers – and not just numbers to Life.

The link between the learning and usefulness of quantities may be viewed the other way round. A student may memorize many of them for examinations and for future clinical application, but which are most profitably learned for the better understanding of the body? Those with most uses? In how many elementary contexts is it helpful to know the concentration of sodium in extracellular fluid? Is that of magnesium as useful? Or of manganese? Such questions of priority are as important for those inclined to overtax their memories unreasonably as they are for the lazy. This book may help both with these decisions and with the learning process itself.

Partly for reasons just indicated, many of my 'numbers' come from textbooks. Working on this text, however, I came increasingly to realize how hard it may be to find what one supposes to be well-known quantities. Textbooks have less and less room for these as other knowledge accumulates, of course, and there is a laudable tendency for concepts to displace quantitative detail. So do not disdain the older books! Diehm (1962) has been a very useful source. Sometimes when a quantitative argument seems frustrated through lack of reliable figures, the solution is to turn it on its head, depart from the natural sequence of calculations, and defer the uncertainties to the end. The reader may spot where I was able to rescue items that way. Only once have I resorted to original data; I am very grateful to Dr Andrew Chappell for dissecting and weighing human muscles for me (see Section 8.4).

I thank also all my colleagues who read portions of draft manuscript or otherwise gave of their time and wisdom, and in particular Dr F. L. Burton, Professor J. V. G. A. Durnin, Dr M. Holmes, Dr O. Holmes, Professor S. Jennett, Dr D. J. Miller, Dr J. D. Morrison, Dr G. L. Smith and Dr N. C. Spurway.

R. F. Burton

How to use this book

Understand the objectives as stated in this Preface; be clear what the book is – and what it is not.

Consider carefully the validity of all assumptions and simplifications.

Much use is made of standard, 'textbook' quantities; ask yourself whether they match those with which you are familiar.

Try guessing answers before calculating, for some are intended to surprise.

The book is primarily about physiology, but a secondary objective is to encourage and facilitate quantitative thinking about the subject. Do not be discouraged if you do not feel yourself to be good at arithmetic, for neither am I. Have some old envelopes handy, though the calculations are designed to minimize the need for such aids. Readers concerned only with the physiology may well choose to use their calculators, or turn straight to the answers at the back of the book.

Do not read too much at a time.

1

Introduction to physiological calculations: approximation and units

My aim in this chapter is to ease progress through those that follow, and generally to lessen the impediments to casual calculation. If some readers find the chapter unnecessary, others may find it particularly educational. The arithmetic in later chapters has been deliberately made simple so that much of it can be done easily in the head (nor is it necessarily wrong to turn straight to the answers at the back of the book!). However, it is useful to be able to cut corners in arithmetic when a calculator is not to hand and guidance is given on how and when to do this. The later part of the chapter is about units. For various reasons these can be extremely troublesome at times, yet the habit of keeping track of them during calculations can be a big help to clear thinking.

1.1 Arithmetic – speed, approximation and error

We are all well drilled in accurate calculation and there is no need to discuss that; what some people are resistant to is the notion that accuracy may sometimes take second place to speed or convenience. Accuracy in physiology is often unattainable anyway, through the inadequacies of data. These points do merit some discussion. Too much initial concern for accuracy and rigour should not be a deterrent to calculation, and those people, especially, who confuse the precision of their calculators with accuracy are urged to cultivate the skills of rough ('back-of-envelope') arithmetic. What is discussed here are these skills, the tolerances implicit in physiological variability and the necessity, at times, of making simplifying assumptions. Thus may one aim of the book be furthered – to be 'an encouragement to quantitative thinking'.

On the matter of approximation, one example should suffice. Consider the following calculation:

$$311/330 \times 480 \times 6.3.$$

An approximate answer is readily obtained as follows:

(nearly 1) × (just under 500) × (just over 6) = slightly under 3000.

The 480 has been rounded up and 6.3 rounded down in a way that should roughly cancel out the resulting errors. As it happens, the error in the whole calculation is only 5%.

When is such imprecision acceptable? Here is another calculation: *In a man of 70 kg a typical mass of muscle is 30 kg: what is that as a percentage?* An answer of 42.86% is arithmetically correct, but absurdly precise, for the weight of muscle is only 'typical', and it cannot easily be measured to that accuracy even with careful dissection. An answer of 43%, even 40%, would seem precise enough.

Note in this example that the two masses are given as round numbers, each one being subject both to variation from person to person and to error in measurement. This implies some freedom for us to change one or other of them slightly here and it so happens that a choice of 28 kg for the weight of muscle, instead of 30 kg, would make the calculation easier. Many of the calculations in this book have been eased for the reader in just this way. That this has not always been done is because of the usefulness of using standard 'textbook' values in order to reinforce these in the student memory. (That textbooks do not always agree on these exactly troubles many students – even when the latter know of the great variability of most physiological measures.)

Appendix A is about logarithms. These feature, for example, in the Nernst and Henderson–Hasselbalch equations. The most important message in Appendix A is that knowledge of the numerical value of only one logarithm (other than of 10, 100, etc.) can be very useful in mental arithmetic, the choice being log 2 (=0.301). It is not that the logarithm of 2 is called for by chance more than others, but rather that one can sometimes tailor one's calculations to utilize it. This is exemplified in later pages, and log 2 also happens to appear naturally in equations in Section 3.7.

Rough answers will often do, but not major errors. Often the easiest mistake to make is in the order of magnitude, i.e. the number of zeros or the position of the decimal point. Here, again, the above method of approximation is useful – as a check on order of magnitude when accurate arithmetic is also required. Other ways of avoiding major errors are discussed elsewhere (see the Section, Units, below).

Wrong answers can obviously be obtained if the basis of a calculation is at fault. However, some degree of simplification is often sensible as a first step in the exploration of a problem. Many of the calculations in this book involve simplifying assumptions and the reader would be wise to reflect on their appropriateness; there is sometimes a thin line between what is inaccurate, but helpful in the privacy of one's thoughts, and what is respectable in print. Gross simplification can indeed be helpful. Thus, the notion that the area of body surface available for heat loss is proportionately less in large mammals than small, is sometimes first approached, not without some validity, in terms of spherical, legless bodies. The word 'model' is useful in such contexts – as a respectable way of acknowledging or emphasizing departures from reality.

1.2 Units

Too often the simplest physiological calculations are hampered by the fact that the various quantities involved are expressed in different systems of units for which the interconversion factors are not immediately to hand. One source of information may give pressures in units of mmHg, and another in cm H_2O, N/m^2 ($N \cdot m^{-2}$) or $dyne/cm^2$ – and two or three diverse values may need to be combined in the calculation. Spontaneity and enthusiasm suffer, and errors in calculation are more likely.

One might therefore advocate a uniform system for physiology and for this book, and in particular the metric *Système International d'Unité* or SI, with its coherent use of kilogram mass, metre and second. However, even if SI units are universally adopted, the older books and journals will remain as sources of quantitative information (and one medical journal has recently abandoned the exclusive use of SI units). In this book I use the units that seem usual in current textbooks and in hospitals, and the reader is not required to struggle with conversion factors. This way, the calculations can serve best their subsidiary function of reinforcing basic knowledge of important quantities such as blood pressures and lung volumes. What is sometimes lost thereby is elegance, as when, in Section 4.10, the law of Laplace, so neat in SI units, is re-expressed in other terms.

Table 1.1 lists some useful conversion factors, but it is not much needed for the calculations in the book. Rather, it is for general reference and as 'an encouragement to (other) quantitative thinking'.

Students often quote quantities without specifying units. All know that units and their interconversions have to be correct, but the benefits of keeping track of units when calculating are not always fully appreciated.

Table 1.1 *Conversion factors for units*

Time		
1 day	1440 min	
	86 400 s	
Distance		
1 m	39.4 inch	
1 foot	**0.305 m**	
1 km	0.621 mile	
1 Ångstrom unit	0.1 nm	
Velocity		
1 mph	**0.447 m/s**	1.609 km/h
Acceleration (gravitational)		
g	**9.807 m/s²**	32.17 ft/s²
Mass		
1 lb	**0.4536 kg**	
Force		
1 N	**1 kg m/s²**	
1 kg-force	**9.807 N**	
1 dyne	**10^{-5} N**	1 g·cm/s²
Energy		
1 J	**1 N·m**	
1 erg	**10^{-7} N m**	1 dyne·cm
1 calorie	**4.1855 J**	
1 m·kg-force (1 kg·m)	**9.807 J**	
Power		
1 watt	**1 J/s**	
Pressure and stress		
1 N/m²	1 Pa	
1 kg force/cm²	**9.807 N/m²**	
1 torr	1 mmHg	
1 mmHg	**133.3 N/m²**	133.3 kPa
750 mmHg	**100 kN/m²**	
1 mmHg	1.36 cmH_2O	
1 cmH_2O	**98.1 N m^{-2}**	

SI units, fundamental or derived, are in **bold** lettering.

Of the examples given below, that of the diffusion coefficient is chosen because the units in which it is expressed may not be intuitively obvious, and also because there is a danger of mixing up rather different kinds of diffusion constant. Each of the following examples is intended to illustrate general points, as well as having specific relevance to physiology.

1.2.1 **The solubility of oxygen** (see also Section 4.2)

The concentration of a oxygen in simple solution, $[O_2]$, increases with the partial pressure, P_{O_2}, and with the solubility coefficient, S_{O_2}:

$$[O_2] = S_{O_2} \cdot P_{O_2}. \tag{1.1}$$

The concentration may be wanted in mmol/l or in ml/l, and the partial pressure in mmHg, kPa or atmospheres. If the chosen units are mmol/l and mmHg, then, for compatibility, the solubility coefficient must be in mmol/l per mmHg. Equation (1.1) may be rewritten in terms just of these units:

$$\text{mmol/l} = \frac{\text{mmol/l}}{\text{mmHg}} \times \text{mmHg}.$$

Treated algebraically the 'mmHg' cancels to give

$$\text{mmol/l} = \text{mmol/l}.$$

The point of spelling this out is that, if one used the wrong form of solubility coefficient and came up with the following instead, the need to think again would be at once apparent:

$$\text{mmol/l} = \frac{\text{ml/l}}{\text{atmosphere}} \times \text{mmHg}.$$

1.2.2 **Dynamic pressure in flowing fluid (SI units)**

A fluid of density ρ is pumped along a tube with velocity v and at pressure P. The dynamic, or total, pressure of the fluid (blood flowing within a vessel perhaps) is given by the following expression, where h is the height to which the fluid is pumped and $\boldsymbol{g}$ is the acceleration due to gravity:

$$(P + \rho g h + \tfrac{1}{2}\rho v^2).$$

The third term, $\frac{1}{2}\rho v^2$, represents the kinetic energy of the fluid. The point of looking at this expression here is that, at first glance, it may be far from obvious that all three terms are compatible in their units. Here they are analysed in SI units:

$$\frac{\text{N}}{\text{m}^2} + \frac{\text{kg}}{\text{m}^3} \times \frac{\text{m}}{\text{s}^2} \times \text{m} + \frac{\text{kg}}{\text{m}^3} \times \frac{\text{m}^2}{\text{s}^2} = \frac{\text{N}}{\text{m}^2} + \frac{\text{kg}}{\text{m}} \times \frac{1}{\text{s}^2} + \frac{\text{kg}}{\text{m}} \times \frac{1}{\text{s}^2}.$$

The apparent disparity is resolved through the fact that $1\ \mathrm{N} = 1\ \mathrm{kg \cdot m/s^2}$. Thus, all three terms can be expressed in units of pressure, such as $\mathrm{N/m^2}$ (= Pa).

1.2.3 **Diffusion**

Suppose that an (uncharged) substance S diffuses from region 1 to region 2 along a diffusion distance d and over an area a. The (uniform) concentrations of S in the two regions are respectively $[\mathrm{S}]_1$ and $[\mathrm{S}]_2$. The rate of diffusion is given by the following equation:

$$\text{rate} = ([\mathrm{S}]_1 - [\mathrm{S}]_2) \times a/d \times D, \tag{1.2}$$

where D is the 'diffusion coefficient'. The appropriate units for D may be found by rearranging the equation and proceeding as follows:

$$\begin{aligned} D &= \frac{\text{rate}}{[\mathrm{S}]_1 - [\mathrm{S}]_2} \times \frac{d}{a} \\ &= \frac{\text{rate}}{\text{concentrations}} \times \frac{\text{distance}}{\text{area}}. \end{aligned} \tag{1.3}$$

The rate of diffusion is the amount of S diffusing per unit of time and concentrations are amounts per unit of volume. Therefore,

$$D = \frac{\text{amount}}{\text{time}} \times \frac{\text{volume}}{\text{amount}} \times \frac{\text{distance}}{\text{area}}.$$

If one chooses to specify distance, area and volume in terms of cm, $\mathrm{cm^2}$ and $\mathrm{cm^3}$, then the expression becomes:

$$\begin{aligned} \text{units for } D &= \frac{\text{amount}}{\text{time}} \times \frac{\mathrm{cm^3}}{\text{amount}} \times \frac{\mathrm{cm}}{\mathrm{cm^2}} \\ &= \frac{\mathrm{cm^2}}{\text{time}}. \end{aligned}$$

Diffusion coefficients, accordingly, are commonly given in units of $\mathrm{cm^2/s}$. Note that it is irrelevant here how the amount of substance is expressed, whether it be in grams, millimoles, etc.

This diffusion coefficient is not to be confused with other diffusion constants with other units. Notably, the diffusion of gases may be treated in terms of partial pressures instead of concentrations. For oxygen, equations (1.1) and (1.2) are combined in the following:

$$\text{rate} = (P_{1O_2} - P_{2O_2}) \times \frac{a}{d} \times (DS_{O_2}). \quad (1.4)$$

(DS_{O_2}) can be regarded as a single entity, a diffusion constant that is clearly distinct from D. Appropriate units, not SI but easily applied, are mmol/mmHg per cm per s. This approach is used in Section 4.3. Only one other point needs to be made for present purposes: diffusion constants may be found elsewhere that are expressed in other combinations of units and which are, or seem to be, of quite different dimensional form. It is therefore especially necessary to keep track of units in this field.

1.2.4 Empirical equations (including allometric equations of Chapter 2)

Typically the above points apply whether a relationship is based on theory or only on experimental observations. Consider, however, the following purely empirical finding. In mammals of different sizes the masses of the kidneys, K, have been found to vary with body mass, M, according to the following allometric equation:

$$K = aM^b. \quad (1.5)$$

The exponent, b, is about 0.85, but subject to uncertainty because of scatter in the data from which it was determined. The proportionality factor, a, has a numerical value of about 0.0073 when the mass units are kg. On the basis of what has been said so far, a should have the units of $\text{kg}/\text{kg}^{0.85}$, i.e. $\text{kg}^{0.15}$. However, one cannot know this until the measurements have been made, and even then the exponent b is uncertain. In these circumstances, correctness in specifying units of a is a matter of form only, awkward, not very useful, and commonly ignored. This sort of problem tends to disappear when relationships are better understood.

2
Energy and metabolism

To feel at home in this field one must be familiar with a variety of measures and units of energy – calories, joules, litres of respired oxygen, slices of bread, and so on. The more books or papers one consults, the more evident does this become. Many of the earlier calculations in this chapter are intended primarily to help with this problem. Amongst other calculations are some to do with the dependence, in mammals generally, of metabolic rate on body size (Section 2.8). Uses and limitations of such allometric scaling are then explored in relation to drug dosage, life span, heart rate, the metabolic cost of sodium transport and the rate of metabolic production of water (Sections 2.9 to 2.12).

Most emphasis is placed here on aerobic metabolism, and on glucose rather than on other metabolic substrates. The reader might like to re-work some of the calculations in terms of, say, anaerobic metabolism or fatty acids.

Amounts of energy are expressed here in terms of both kilocalories and kilojoules. When pairs of such figures (i.e. kcal and kJ) are chosen for their convenience, or as round numbers, then they are not always exactly equivalent. Calculations on energy use are also to be found elsewhere (Sections 3.7, 5.8, 8.4 and 8.5).

2.1 Measures of energy

The variety of units in which energy may be expressed does not help casual quantitative thinking. Most of this chapter makes use of kilocalories (kcal) with kilojoules (kJ) in brackets. As shown in Table 1.1, 1 kcal is 4.2 kJ, so that 1 kJ is 0.24 kcal.

By definition, it takes 1 kcal to raise the temperature of 1 kg of water by one degree Celsius (strictly, from 15 °C to 16 °C). It takes about

0.8 kcal to raise the temperature of 1 kg of human tissue by the same amount. (In other words, the specific heat capacity of the body is 0.8 kcal/kg per degree Celsius).

2.1.1 **How many kilocalories are required to raise the temperature of a 100 kg person by one degree Celsius?**

To put this answer into a context, suppose that that person has a basal metabolic rate of 1900 kcal/day (8000 kJ/day). This is equivalent to 80 kcal/h (333 kJ/h), enough to raise the temperature by one degree per hour if there were no heat loss.

Another unit of energy is the metre·kg-force: 1 mg·kg-force is 9.81 J. (Although this may often be appropriately rounded up to 10 J, it is as well not to lose sight of the fact that the 9.81, as m/s^2, represents $\boldsymbol{g}$, the acceleration due to gravity.)

2.1.2 **In accordance with the last paragraph, how much energy, in (a) kJ and (b) kcal, is required to raise a 100 kg person by 10 m?**

The answers to 2.1.2 make no allowance for inefficiency, e.g. for heat generation within the body of the person doing the lifting (see below).

In this context of work against gravity, joules have an obvious advantage over calories, and later calculations on jumping and lifting (Section 8.4) are framed in terms of joules. In quantifying energy utilization in physiology, however, one may choose to think in terms neither of joules nor calories, but of oxygen consumption or use of energy sources such as glucose and fat.

The metabolic oxidation of glucose may be described by the following overall equation:

$$C_6H_{12}O_6 + 6O_2 = 6CO_2 + 6H_2O. \qquad [2.1]$$

From this it is evident that the complete oxidation of 1 mol of glucose to carbon dioxide and water utilizes 6 mol of oxygen and yields 6 mol of carbon dioxide. Also, since 1 mol of oxygen at standard temperature and pressure (STP) occupies 22.4 l, 1 mol of glucose requires 134.4 l of oxygen. Table 2.1 gives some approximate conversion factors for glucose in accordance with these values – and corresponding factors for fat and protein. The energy data are as determined by combustion in a bomb calorimeter, except that those for protein allow for the energy content

Table 2.1 *Energy content and oxygen consumption for glucose, fat and protein*

	Glucose	Fat	Protein	'Body'
kcal/g	3.8	9.3	4.2	
kJ/g	16	39	18	
l oxygen/g	0.75	2.0	1.0	
kcal/l oxygen	5.1	4.7	4.2	4.6
kJ/l oxygen	21	20	18	19
kcal/mol oxygen	114	105	94	103
kJ/mol oxygen	479	440	393	430
Respiratory quotient	1.00	0.70	0.80	0.80

Energy content of glucose, fat and protein. Also shown are the amounts of oxygen required for complete combustion or metabolism of each, the amounts of energy released per l and mol of oxygen, and the respiratory quotient (the ratio of carbon dioxide molecules released to oxygen molecules consumed). The column headed Body is for a somewhat arbitrary combination of carbohydrate, protein and fat such as might be catabolised in a fairly typical human body.

of the nitrogenous excretory products and are thus somewhat lower. Polysaccharides yield about 4.1 kcal/g (17 kJ/g), so slightly more for a given mass than glucose does. In the last column of Table 2.1 is a somewhat arbitrary combination of carbohydrate, protein and fat: these values are for use in rough calculations to do with a typical human body.

2.2 Adenosine triphosphate and metabolic efficiency

The immediate source of energy for much physiological work is ATP. It hydrolyses thus:

$$\text{ATP} + \text{H}_2\text{O} = \text{ADP} + \text{inorganic phosphate}. \qquad [2.2]$$

The free-energy change for the hydrolysis of ATP to ADP (ΔG_{ATP}) is usually estimated for intracellular conditions as -10 to -13 kcal/mol of ATP (-42 to -54 kJ/mol).

Much of our ATP is generated, by reversal of reaction [2.2], in the course of the oxidative metabolism of glucose (reaction [2.1]). Typically, 1 mol (180 g) of glucose yields 38 mol of ATP.

2.2.1 **If the aerobic metabolism of 1 mol of glucose yields 38 mol of ATP and ΔG_{ATP} is −11 kcal/mol (−46 kJ/mol), how much energy is available from 1 mol of glucose via the hydrolysis of ATP to ADP?**

Burning 1 mol of glucose in a bomb calorimeter yields substantially more energy than this – about 686 kcal (2870 kJ). (This is in accordance with Table 2.1, for the molecular weight of glucose is 180.) The discrepancy between the two values may be expressed in terms of the 'efficiency' of energy transfer. This is simply the answer to question 2.2.1 divided by the full amount of energy that is available from 1 mol of glucose on complete combustion.

2.2.2 **What is the efficiency of energy transfer from glucose to ATP?**

In the light of this result, it is clear that one cannot equate, using Table 2.1, the amount of mechanical work done by a muscle, for example, and the energy from the glucose it utilizes. There are, in any case, further losses of energy as heat in the contraction process, so that the overall efficiency, defined as mechanical work divided by energy input, is even lower (Sections 3.7 and 8.5). The kidneys are also extremely inefficient inasmuch as they use far more energy than the thermodynamic minimum required to produce the urine (Section 5.8).

In a mammal, in contrast to a fish, such energy losses are not all waste, for heat is needed to maintain body temperature. The next Section deals with a very minor aspect of temperature regulation.

2.3 Cold drinks, hot drinks

I can comfortably take tea into my mouth at a temperature of 63 °C. A hotter drink needs to be cooled by sipping. The comfort of a warm drink in cold weather, or of a cold drink in hot weather, is obvious, but the effect on body temperature requires calculation.

As noted in Section 2.1, it takes 1 kcal to raise the temperature of 1 kg of water by one degree Celsius, by definition, and 0.8 kcal to do the same to 1 kg of the body. In the next calculation, ignore this difference in specific heats.

2.3.1 **A woman weighing 60 kg drinks 600 ml (0.6 kg) of water. The temperature of the water is 25 °C above or below her mean body temperature. By how much would the latter change?**

Clearly, the warm drink would do little for severe hypothermia, and the cold drink would do little for a fever. However, even the small change in temperature that has been calculated is much greater than may be needed to evoke shivering or sweating by stimulation of the hypothalamus.

2.3.2 **How much heat, in kcal, must be gained or lost to bring that 0.6 kg of water drunk by the woman to body temperature?**

We may try to give this answer more meaning by placing it in the context of metabolic heat production. Let us therefore postulate that the woman's metabolic rate happens to be 100 kcal/h (2400 kcal/day, 10 000 kJ/day).

2.3.3 **To how many hours-worth of metabolism does the previous answer correspond?**

The heat of hot food and drink is not utilized to perform physiological work and is not in that sense equivalent to metabolic energy. However, hot food or drink does spare metabolic energy when the body is too cold.

2.4 Oxygen and glucose in blood

The oxygen content of blood is usually expressed in terms of volumes. In fully oxygenated human arterial blood it is about 200 ml/l (20 volumes %). This may not relate easily to concentrations of other metabolites since these are usually expressed in other units. Thus, the concentration of glucose is often given in units of mg/100 ml, the value being about 90 mg/100 ml in capillary blood. Given that a glucose molecule requires six oxygen molecules for its complete oxidation (reaction [2.1]), one might wish to know how the concentrations of the two compare when both are expressed as mmol/l. One would then know, for example, whether there is enough oxygen in a given volume of arterial blood to oxidize completely all the glucose that is present.

2.4.1 **What is the above concentration of oxygen, 200 ml/l, re-expressed as mmol/l?** (1 mmol occupies 22.4 ml at STP)

2.4.2 **What is the above concentration of glucose, 90 mg/100 ml, re-expressed as mmol/l?** (The molecular weight of glucose is 180)

It is now clear whether or not there is enough oxygen in the blood to oxidize all the glucose completely – 6 mmol of oxygen for each 1 mmol of glucose. Erythrocytes do not themselves metabolize glucose aerobically, but they do consume it anaerobically, with the release of lactate into the plasma, at a rate of 1.5–2 mmol/h for each litre of cells.

2.4.3 **For blood containing 45% erythrocytes, what is their rate of glucose consumption in mmol/l of blood per hour when each litre of cells utilizes it at a rate of 2 mmol/h?**

2.4.4 **For a total blood volume of 5 l, what then is the rate of glucose consumption by the erythrocytes in (a) mmol/day and (b) g/day?**

Glucose consumption is thus substantial, but much of the lactate produced by the erythrocytes is re-converted to glucose in the liver.

2.5 Basal metabolic rate

The basal metabolic rate (BMR) of a 70 kg man is usually within the range of 1200–2100 kcal/day (5000–9000 kJ/day), depending on such factors as fat content and age. Such figures may be re-expressed, approximately, in terms of oxygen consumption, assuming aerobic metabolism. The relationship between heat production and respiratory exchanges was first measured by Lavoisier and Laplace in 1780, using guinea-pigs. According to the last column of Table 2.1, 1 l of oxygen is equivalent to about 4.8 kcal (20 kJ).

2.5.1 **If a man's basal metabolic rate is 1700 kcal/day (7000 kJ/day), what is his likely rate of oxygen consumption under basal conditions, in l/day and l/min?**

A convenient round value to remember is 0.25 l/min.

It is not hard to visualize one's daily food intake, but it includes water, the bubbles in bread, and indigestible cellulose, and one's metabolic rate is generally well above basal. It may, therefore, be helpful to think of the above basal metabolic rate in terms of purer fuels.

2.5.2 **For the same metabolic rate of 1700 kcal/day (7000 kJ/day), what would the daily rates of fuel consumption be in grams, if the fuel were (a) entirely glucose and (b) entirely fat?** (see Table 2.1)

Should you wish to compare the power output of the body with that of a light bulb, recall that 1 watt = 1 J/s and that 1 day = 86 400 s.

2.5.3 **If a person's metabolic rate happens to be 8640 kJ/day (2064 kcal/day), what is that value in watts?**

2.6 Oxygen in a small dark cell

You are thrown into a small dark room. The door clangs shut and you fear that there are no holes or cracks for ventilation. How long will your air supply last? Reaching up, you easily touch the ceiling – seven feet high, perhaps? The floor is about six by seven feet, so what is the volume, allowing for your own volume? And in litres? There is no point in accuracy here, but a two-metre cube should be nearly right, i.e. 8000 l. About one-fifth of that should be oxygen, i.e. 1600 l. You recall that basal oxygen consumption is typically about 0.25 l/min, or 15 l/h, so perhaps a convenient 16 l/h would do for your calculation if you keep still. Never mind corrections for temperature and pressure!

2.6.1 **Assuming (for the calculation only!) a steady rate of consumption, how long would your oxygen last?**

2.6.2 **Another worrying thought: assuming that you release carbon dioxide at the same rate as you use up oxygen, what would be the percentage carbon dioxide content of the air when half that time had elapsed?**

2.6.3 **What is that answer when expressed as P_{CO_2} in mmHg? (Assume a barometric pressure of 760 mmHg.)**

Naturally enough in the circumstances, a few complicating factors have been ignored. You would die before all the oxygen had gone. The rate of ventilation would rise with rising P_{CO_2} and falling P_{O_2}, and with it the work of breathing – hence also the total oxygen consumption which was assumed to be constant. Your body would accumulate excess carbon dioxide and not release all of it to the room.

2.7 Energy costs of walking, and of being a student

First a calculation of possible interest to slimmers, though formulated in terms of the physiologist's standard, non-obese, man. The point is to express the energetic cost of walking in terms of a familiar fuel. Glucose has been chosen, although skeletal muscle actually takes up free fatty acids, acetoacetate and β-hydroxybutyrate in preference to glucose.

A 70 kg man sitting quietly expends, say, 90 kcal/h (377 kJ/h, 2160 kcal/d, 9048 kJ/day). His energy consumption walking on level tarmac at 4.5 km/h (2.8 m.p.h.) would be about 240 kcal/h (1000 kJ/h). Glucose supplies 3.8 kcal/g (16 kJ/g) (Table 2.1).

2.7.1 **What would be the extra cost of walking 4.5 km (2.8 miles) at that speed as compared with simply sitting and thinking about it? Express the answer (a) in kcal or kJ and hence (b) in terms of grams of glucose.**

Since one does not ordinarily subsist on glucose alone, here is the answer re-expressed, roughly, in terms of other foods: 65 g of bread, 21 g of butter, 28 g of potato chips, 1150 g of cucumber.

The work of walking has been measured for different speeds, different terrains, etc., but calculating it from first principles is not easy, since that requires too much knowledge of what the individual muscles are doing. What can be easily calculated is the minimum extra cost of walking uphill, above that required for walking on the level, i.e. the external, mechanical work. This may then be compared with actual, measured, energy expenditure. As already noted in relation to question 2.1.2, the mechanical work done in moving a body upwards is 9.8 J/(kg · m). This value, conveniently close to 10 and with its reminder of '*g*', is used below, rather than the calorie equivalent, 2.3 cal/(kg · m).

It has been found that walking on an upward slope requires an extra energy expenditure of about 7 cal (30 J) per kg of body mass for each metre that one ascends, this being additional to the cost of walking the same horizontal distance on the level. Values similar to this have been obtained for a variety of different kinds of mammal.

2.7.2 **If the actual work of walking up a slope, in excess of the work done in walking on the level, is 30 J/(kg · m), and the (minimum) mechanical work done is 9.8 J/(kg · m), what is the percentage efficiency?**

A lower value for efficiency would, of course, be obtained if the total work of walking were considered.

The energy cost of the ascent is much the same for mammals of different size, when expressed per kilogramme. That small mammals climb and run uphill with more apparent ease has to do with their generally higher metabolic rates (Section 2.8); the extra work against gravity is proportionately less.

To estimate one's daily energy expenditure, one may look up in tables the energy costs of one's principal activities and add them up, with due regard to body mass, sex and so on. That is not an appropriate exercise here, but we can try to relate data we have already to daily metabolic rate.

Let us take the case of a fairly typical university student expending 2610 kcal/day (10 925 kJ/day). As an undoubted simplification, to see where it gets us, we may take the student's day as being divided between two activities, quantitatively as above – walking with an energy expenditure of 240 kcal/h (1000 kJ/h) (or doing anything else of comparable energy cost) and resting or studying with an energy expenditure of 90 kcal/h (377 kJ/h). From this information we can estimate how many hours of the day (x) is spent in walking, for (working in terms of kcal)

$$x \times 240 + (24 - x) \times 90 = 2610. \qquad (2.1)$$

Hence,

$$150x = 450.$$

2.7.3 **How much of the day does this student spend in walking, or in activities of equivalent energy requirement?**

Is this a credible result? How might you improve on this method? It would not work for such people as farmers, coal miners and forestry workers. They might use up to about 4600 kcal/day (19 000 kJ/day).

2.8 Basal metabolic rate in relation to body size

2.8.1 **A particular 70 kg man has a basal metabolic rate (BMR) of 1700 kcal/day (7000 kJ/day). A mouse weighing 30 g has a BMR of about 4.8 kcal/day (20 kJ/day). In each case, what is the BMR expressed per kilogram of body mass (i.e. kcal/(kg·day) or kJ/(kg·day))?**

Expressed in this way, per unit mass, the BMR is known as the *specific BMR*. The general bounciness of small dogs compared with big ones is well known – as is the more sedate behaviour of elephants as compared with mice. Such differences apply to basal metabolism as well as to general activity. As long ago as 1839 the assumption was made (by Sarrus and Rameaux) that metabolic rate in mammals of different mass (M) is proportional, not to M itself, but to $M^{2/3}$. Since that time there has been much measurement and theorizing and we will return later to the correctness or otherwise of that two-thirds exponent. First, however, let us first see why, in mammals, the exponent cannot be 1.0. (With metabolic rate proportional to $M^{1.0}$, the specific metabolic rate is constant.)

Recall now a point of solid geometry, that for cubes of side length L the volume is proportional to L^3 and the surface area is proportional to L^2. Therefore, the ratio of surface area to volume is proportional to $L^{2/3}$. This is true of any three-dimensional bodies of constant shape. Moreover, if the density is constant, the ratio of surface area to mass (M) is proportional to $M^{2/3}$.

Imagine two mammals that differ in body mass by a factor of 1000 but which are otherwise similar in almost every respect, including shape and fur thickness. Their surface areas must differ by a factor of 100. They are postulated as both having the same metabolic rate per unit body mass (in accordance with an exponent of 1.0), so that the absolute metabolic rate of the larger animal is 1000-fold that of the smaller. Metabolic heat is assumed to be lost to the environment at a rate that is proportional to the surface area and to the difference in temperature between the body and the environment. (Heat loss from the body is not really that simple, especially if much occurs by evaporation.)

2.8.2 **If the smaller mammal has a body temperature of 37 °C and the temperature of the environment is 17 °C, what must the body temperature of the larger mammal be for it to remain in heat balance?**

If the assumptions behind this calculation are over-simple, the conclusion is clear enough!

Data on BMR in mammals are roughly predictable from the following equation:

$$\text{BMR} = a \cdot M^b, \tag{2.2}$$

where M is again body mass and a and b are constants. In logarithmic form, the equation becomes:

$$\log \text{BMR} = \log a + b \log M. \tag{2.3}$$

The two equations do not represent an exact rule, for both individual animals and individual species may diverge from them. Nevertheless, the average data for different species fall neatly around a straight line when BMR and M are plotted on a graph as logarithms. It is important to realize that the equations are regression equations that have been fitted to such data; they are descriptive and do not necessarily have a clear theoretical basis. Much has been written elsewhere on their validity and interpretation.

For mammals as diverse in size as shrews and whales, the value of b has been found to be close to 0.75 rather than to 2/3. The corresponding proportionality coefficient, a, is equal to about 70 for kcal/(kg·day), or 293 for kJ/(kg·day).

2.8.3 **For comparison with data in 2.8.1, what is the BMR predicted from equation (2.2) for a 70 kg man?** ($70^{0.75} = 24.2$)

2.8.4 **A typical BMR for a male child of 10 kg is 600 kcal/day (2500 kJ/day). For comparison, what value is predicted from equation (2.2)?** ($10^{0.75} = 5.6$)

What one might regard as reasonable agreement when making a broad survey of metabolic rates in different mammals will evidently not do in the context of predicting exact rates in men and boys (or women and girls) of different ages. There are age-dependent factors other than body mass.

2.8.5 **For use later (in 2.10.2), what is the *specific* BMR of a blue whale of 100 000 kg?** ($100\,000^{0.75} = 5623$)

Comparison of the answers to questions 2.8.5 and 2.8.1 suggests one reason why blue whales can remain submerged much longer than we (or even dolphins) can.

Human metabolic rates are sometimes expressed per unit area of body surface, but this is usually estimated from a more complicated formula than one utilizing simply $M^{2/3}$.

Many other physiological variables scale with body size in accordance with equations such as (2.2) and (2.3). These are known as allometric equations. Examples of such variables include kidney mass (see equation 1.5), cardiac output and glomerular filtration rate. Other variables, such as arterial blood pressure, seem to be independent of body mass.

2.9 Drug dosage and body size

If the dosage of a drug required to produce a given effect is known for a small laboratory animal, how does one allow for the size difference in order to estimate the dosage appropriate to a human subject? Should the dose be scaled in proportion to body mass, or should it be related to metabolic rate? The first might be appropriate if what matters is the maximum concentration following rapid dispersal of the drug within a particular body compartment. Scaling in relation to metabolic rate makes sense if it is supposed that drug clearance, by liver, kidneys or whatever, is related to general metabolism and renal filtration. There is no exact and general answer to the question, and specific differences in both drug response and drug metabolism may be more important anyway. However, one can at least calculate the difference between the two approaches.

Consider the 30 g mouse and the 70 kg man of 2.8.1. Suppose that a 'suitable' dosage of a drug for the mouse is found to be A units. If the dose for the man is estimated by scaling up in proportion to body mass, then it is $A \times 70/0.03 = 2333A$ units. The basal metabolic rates of mouse and man were given as 4.8 kcal/day (20 kJ/day) and 1700 kcal/day (7000 kJ/day), respectively.

2.9.1 **Scaled in proportion to BMR, what would be the appropriate dosage for a 70 kg man?**

2.9.2 **As a round number, what is the ratio of the two estimated doses?**

An elephant was once given a dose of LSD based on the known effects of the drug on cats and humans and scaled up on the basis of body mass. The elephant died dramatically and unpleasantly.

Scaling of drug doses on the basis of body surface area has sometimes been advocated on the assumption that BMR is related to surface area (Section 2.8). Because of possible interspecific differences in drug response and metabolism, any such scaling can only be regarded as a useful preliminary to empirical trials.

2.10 Further aspects of allometry – life-span and the heart

An idea widely current is that our natural life-span extends, roughly speaking, to a certain number of heart beats. Presumably few take it more seriously than they do our three score years and ten – especially amongst those who seek to prolong their life by exercising heart and limbs! Nevertheless, we may usefully explore the matter in the context of mammals in general and thereby meet further aspects of allometry and its interpretation.

Equation (2.2) relates metabolic rate to body mass in different species of mammal. Here are two more such regression equations. The first relates the frequency of heart beat at rest to body mass (M, in kg) and the second relates life-span in captivity to body mass.

$$\text{heart rate} = 241M^{-0.25} \text{ beats/min.} \tag{2.4}$$

$$\begin{aligned}\text{life-span} &= 11.8M^{+0.20} \text{ years}\\ &= 6 \times 10^6 M^{+0.20} \text{ min.}\end{aligned} \tag{2.5}$$

In the following calculation ignore the fact that equation (2.4) relates to resting, rather than average, heart rates.

2.10.1 **According to these equations, what is the relationship between body mass and the average number of heartbeats in a life time (heart rate × life-span)?**

What concerns us here is whether or not the exponent is zero, for, if it is, then the number of heart beats in a lifetime (of leisure) is about the same in large and small mammals. In fact the exponent is not zero, but it is small. Whether the discrepancy is statistically significant is another matter, for the two equations are but regression equations based on varied data. Let us consider, not the statistics, but how much difference the discrepancy actually makes. This we can explore in terms of the ratio $M^{-0.05}/M^0$ ($=M^{-0.05}$). If this varies little enough with M, then 'beats/lifetime' is nearly independent of M. For a man weighing 70 kg, $M^{-0.05}$ is equal to 0.8. What about smaller mammals?

2.10.2 **To take the easiest case, what is $M^{-0.05}$ in a 1 kg mammal?**

The fact that $M^{-0.05}$ is not very different in the two mammals implies a similarity in average 'beats/lifetime' that would be hard to disprove. It is not to be taken too seriously.

The next two questions are obvious ones to ask in relation to what we know about ourselves, but they are posed more in order to make a point about the allometric regression equations.

2.10.3 **For a 70 kg person, what is the resting heart rate as predicted from equation (2.4)?** ($70^{-0.25} = 0.35$)

2.10.4 **For a 70 kg person, what is the length of life, in years, predicted from equation (2.5)?** ($70^{+0.20} = 2.3$)

If either answer is unrealistic, remember that the equations represent regression lines through collections of data showing considerable scatter. This is not just due to measurement error, but reflects the considerable anatomical and physiological differences amongst species even of the same body size. Longevity happens to be better predicted if brain size, as well as body mass, is taken into account. Allometric equations summarize general trends and are not exactly predictive.

2.11 The contribution of sodium transport to metabolic rate

Sodium is continuously entering cells – in association with action potentials and sundry co-transport mechanisms for example – and it is continuously being baled out again by Na, K ATPase. Typically, the hydrolysis of one ATP molecule powers the transport of three sodium ions. Just how much of the total metabolic rate of the resting human body is normally devoted to sodium transport is hard to say, but a range of 20–45% has been suggested, with 20% as a recent estimate. It is interesting to explore the matter in the context of the considerable size dependence of metabolic rate in different species of mammal (Section 2.8). Given that cell sizes, body temperature and ionic concentrations vary little from one species to another, one might perhaps suppose that the rate of sodium pumping is about the same *per kg of body mass* in them all. At any rate, let us try out that assumption on man and whale.

2.11.1 **Consider the man described in equation 2.8.1 for whom specific BMR was calculated as 24 kcal/(kg·day) (100 kJ/(kg·day)). Assume that 20% of the BMR is devoted to sodium transport. How much is that in kcal/(kg·day) or kJ/(kg·day)?**

2.11.2 **Consider now the blue whale of 100 000 kg for which specific BMR was calculated (in** 2.8.5) **as 3.9 kcal/(kg·day) (16.5 kJ/(kg·day)). Assuming that just as much of the specific BMR is associated with sodium transport in terms of kcal/(kg·day) or kJ/(kg·day) as in the man, what is that as a percentage of the total?**

Could something be wrong with the assumptions? Perhaps much less energy is used for sodium transport in the resting human body than was estimated. Perhaps rates of sodium transport in different species are more nearly proportional to metabolic rate than to body mass. In the case of the kidneys, both the metabolic rate and the rate of sodium transport are closely linked to glomerular filtration rate (Section 5.8) and that scales with body mass much as does BMR.

2.12 Production of metabolic water in human and mouse

Water is produced in the catabolism of carbohydrates, proteins and fats and makes a significant contribution to body water balance. Just how much is produced in a day depends on the metabolic rate and on the proportions of carbohydrates, proteins and fats being metabolized, but a rough value is easily calculated. Disregarding proteins and fats, let us consider only the oxidation of glucose in reaction [2.1] in which one water molecule is produced for each molecule of oxygen consumed. For our daily oxygen consumption, let us take the basal value used earlier of 0.25 ml/min = 15 l/h = 360 l/day. Since 1 mol of oxygen occupies 22.4 l (at STP), 360 l/day is equivalent to 16 mol/day, so that the rate of water production is 16 mol/day too. The molecular weight of water is 18.

2.12.1 **Under those circumstances, what is the rate of metabolic water production in ml/day?**

That answer corresponds to a particular rate of basal metabolism. A value commonly given for a typical active person is 400 ml/day. As a round number it is also the rate of water loss in expired air (Section 4.6).

Now we come to the mouse. Basal metabolic rates per kg of body (specific BMRs) were calculated in question 2.8.1 as 24 kcal/(kg·day) (100 kJ/(kg·day)) for the man and 160 kcal/(kg·day) (667 kJ/(kg·day)) for the mouse. The rate for the mouse is thus 6.7-fold higher than that for the man. Presumably the mouse also produces metabolic water roughly seven times more rapidly than we do.

2.12.2 **Assuming that we actually produce metabolic water at a rate**

of about 400 ml/day, how fast would we produce it if we had the specific metabolic rate of a mouse?

For comparison, we typically take in some 1.5–3 l of water per day as food and drink. Some desert rodents do not need to drink at all, even when subsisting on dry seeds.

3
The cardiovascular system

In this chapter we progress from blood, through blood vessels, to heart. We start with the erythrocytes (Sections 3.1 and 3.2): circulatory haematocrit, maximum and optimum haematocrits, maximum osmotic swelling, lack of nucleus. Moving on to the blood vessels, we consider peripheral resistances (at rest and in exercise, in the systemic circulation and in the lungs) and then some aspects of blood flow and gas exchange in the contexts of blood vessel diameters, body size and exercise. In Section 3.5 the Law of Laplace is applied to arteriolar smooth muscle. The importance of the Frank–Starling mechanism in the matching of right and left cardiac outputs is stressed by means of a rather obvious calculation that is applicable to other homeostatic balances. Finally, in Section 3.7 I offer calculations on the work of the heart.

Two well-known relationships might seem to be under-exploited in this chapter – Poiseuille's equation, and the Law of Laplace. Here is Poiseuille's equation for the rate of flow of fluid in a cylindrical tube due to a difference in pressure between the two ends:

$$\begin{aligned} \text{flow} &\propto \text{pressure difference/resistance} \\ &= \text{presssure difference} \times \frac{\text{radius}^4}{\text{viscosity} \times \text{length}}. \end{aligned} \tag{3.1}$$

Despite the undoubted importance of the formula in elementary teaching, the realities of blood flow are more complicated; flow may be pulsatile or turbulent, viscosities vary with circumstances, the fourth power does not always apply. Thus it is that the full equation does not reappear as such, although components of it do. The Law of Laplace is applied here only to arterioles. As algebra, the formula also illuminates cardiac function, but the heart has too complicated a shape for simple quantitative treatment.

For a discussion of the renal regulation of blood pressure, see Chapter 5.
For a calculation on Purkinje fibres, see Chapter 8.

3.1 Erythrocytes and haematocrit (packed cell volume)

Erythrocytes make up about 36–50% of human blood, averaging about 46% in men and 41% in women. The proportion in a blood sample may be estimated by centrifugation in a cylindrical tube (e.g. a Wintrobe tube) followed by measurement of the length of the column of packed erythrocytes relative to that of the whole column of blood. The resulting 'haematocrit' or 'packed cell volume' (PCV) may be expressed as a percentage or decimal fraction. Some 3–8% of plasma remains occluded amongst the erythrocytes after centrifugation and if a correction is applied for this, the averages reduce to 44% for men and 39% for women.

An estimate based on a blood sample from a large vessel may be called the 'central haematocrit' to distinguish it from the 'mean circulatory haematocrit' which, as explained below, is slightly less. The mean circulatory haematocrit is the percentage of erythrocytes in the blood calculated from measurements of the body's content of plasma and circulating erythrocytes.

3.1.1 **Using typical textbook values for plasma, erythrocyte and blood volumes (sex unspecified) of 3 l, 2 l and 5 l respectively, what is the mean circulatory haematocrit?**

The answer lies between the above corrected haematocrits of 44% for men and 39% for women (of which the mean is 41.5%). For mnemonic purposes this is excellent agreement and one should not ask more of round numbers and averages from varied sources. However, the mean circulatory haematocrit should actually work out less than the corrected central haematocrit. This is because the erythrocyte content of the blood in the capillaries and arterioles tends to be reduced by axial streaming of the corpuscles. The discrepancy averages 9%, so that the mean circulatory haematocrit is about 0.91 times the corrected central haematocrit.

3.1.2 **If the corrected central haematocrit is 41.5% (the mean for men and women given above), what is the mean circulatory haematocrit likely to be?**

A more satisfying correspondence with the answer to question 3.1.1 is

Table 3.1 *The diameter, greatest thickness and volume of a typical human erythrocyte*

Diameter	8.3 μm
Greatest thickness	2.4 μm
Volume	84 μm^3

achieved (to three significant figures) if one starts with a corrected central haematocrit of 44%, the average given above for men.

However the haematocrit is defined, there is little distance between the corpuscles. One may wonder, therefore, how much more closely they could be packed without distortion. (The effect of varying haematocrit on viscosity is considered in Section 3.2.) To explore this we need suitable dimensions for a human erythrocyte and these are given in Table 3.1. (The diameter of 8.3 μm is for circulating blood; in blood smears it is about 7.5 μm and in tissue sections, where it is a useful indicator of scale, it is typically nearer 7 μm.)

To calculate the percentage haematocrit corresponding to the closest possible packing of undistorted erythrocytes, one may think of the latter as oriented in stacked sheets; in each sheet the cells are laid out touching each other like coins lying flat on a table (Fig. 3.1). Each erythrocyte may then be regarded as occupying an imaginary hexagonal box that shares each side with its neighbours. The erythrocytes in their boxes are thus close-packed like bee grubs in their honeycomb cells. The required answer is given by the percentage volume of any one box that is occupied by its erythrocyte. The area of a hexagon enclosing a circle of radius R is $3.47R^2$ (3.47 being $6 \div \sqrt{3}$). For the erythrocyte of Table 3.1, this is 60 μm^2.

3.1.3 **What percentage haematocrit corresponds to the closest possible packing of erythrocytes without distortion?**

Not surprisingly, the answer is greater than any normal haematocrit value. In polycythemia, however, haematocrits may attain 70%, and this indicates that the cells cannot then be of normal shape. Obviously, erythrocytes are much more closely packed after centrifugation (with only a little plasma amongst them) and must become considerably distorted in the process.

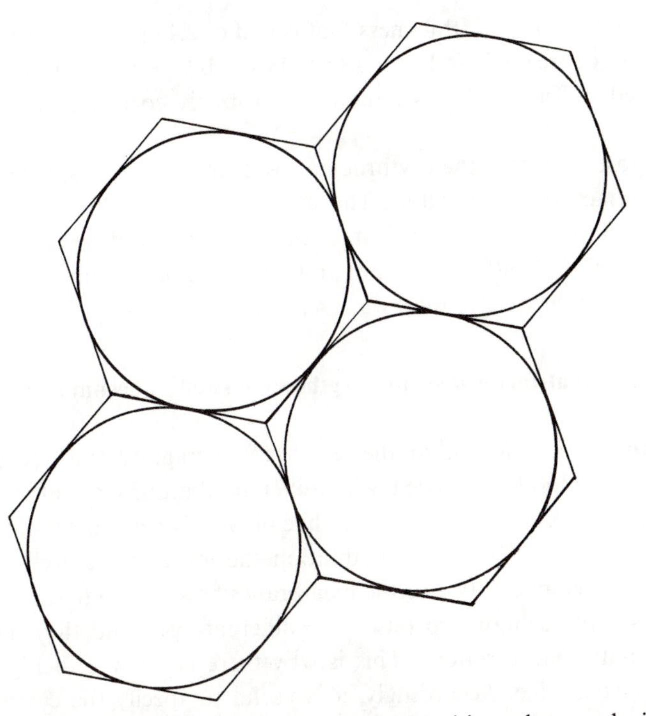

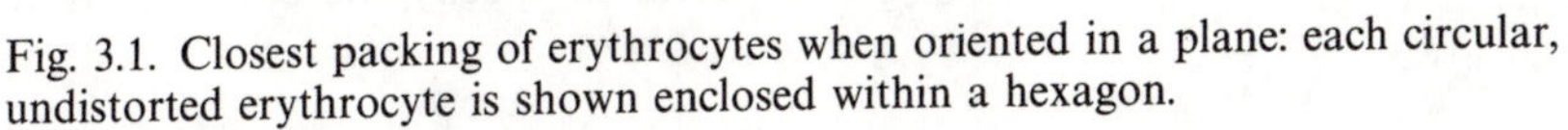
Fig. 3.1. Closest packing of erythrocytes when oriented in a plane: each circular, undistorted erythrocyte is shown enclosed within a hexagon.

Perhaps the volume given for the erythrocyte (84 μm^3), or a value like it, is already familiar. If not, it is readily estimated from two other well-known quantities, as in the next calculation.

3.1.4 **In men, the average red cell count in peripheral venous blood is about 5.4 million per mm^3. If the corresponding (corrected) haematocrit is 44%, what is the average erythrocyte volume ('mean corpuscular volume')?** ($1\ mm^3 = 10^9\ \mu m^3$)

A feature of the shape of the erythrocyte is that swelling can occur without increase in surface area, e.g. by osmosis when the erythrocyte is placed in a dilute salt solution. Only when the corpuscle has become spherical is there any tension on the cell membrane that could lead to haemolysis. The associated increase in volume can be calculated for the erythrocyte considered in Table 3.1, provided that the surface area is also known. This can be estimated closely enough by treating the biconcave disc as a cylindrical disc of only half the thickness. For the diameter of 8.3 μm (as in Table 3.1) and a thickness that is half of 2.4 μm, the cylindrical disc has a surface area of $2\pi(4.15^2 + 4.15 \times 1.2) = 140\ \mu m^2$. For the erythrocyte described in Table 3.1 this estimate is probably correct to within a few percent.

Suppose now that the erythrocyte swells to become a sphere with the same surface area of $140\ \mu m^2$. The surface area of a sphere of diameter d is πd^2 and its volume is $\pi d^3/6$. It follows that the swollen erythrocyte has a diameter of $\sqrt{(140/\pi)} = 6.7\ \mu m$ and a volume of $156\ \mu m^3$. The original volume, as a biconcave disc, was $84\ \mu m^3$.

3.1.5 **By what factor does this erythrocyte swell in becoming spherical?**

This answer is relevant to the 'erythrocyte fragility test'. In this test, samples of blood are mixed with different dilutions of saline and the subsequent occurrence of osmotic haemolysis is related to salt concentration (Fig. 3.2). With sufficient dilution, the erythrocytes swell to become spheres – then haemolyse as the membranes become stretched. Any given blood sample contains corpuscles of different ages, and they haemolyse at different concentrations. This is why there is a curve in Fig. 3.2 and not a vertical line. Accordingly, it is useful to specify the concentration

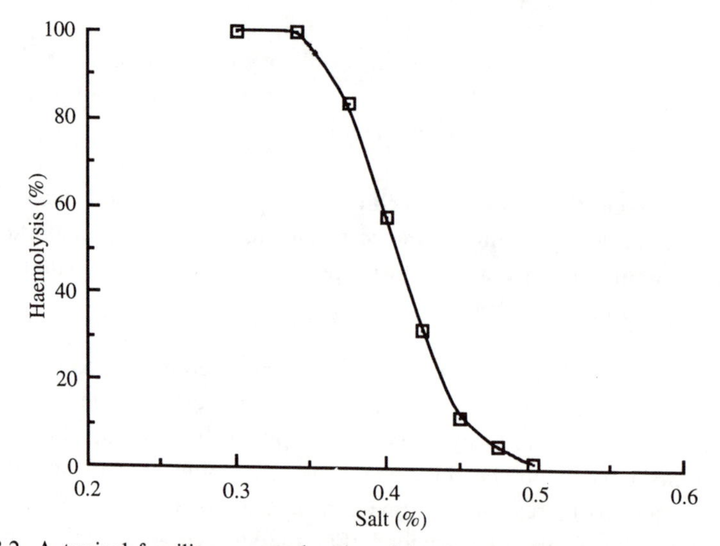

Fig. 3.2. A typical fragility curve, showing the relationship between percentage haemolysis and the concentration of the dilute NaCl solution in which the erythrocytes are suspended. The normal concentration is 0.9%.

at which 50% of corpuscles haemolyse. This is typically between about 0.37 and 0.44% NaCl. Haemolysis occurs at higher concentrations when the corpuscles start more spherical, as in hereditary spherocytosis.

Let us now estimate the concentration at which the erythrocyte of question 3.1.5 would haemolyse. As an approximation, the volume of a corpuscle is inversely proportional to salt concentration, but a more exact relationship is obtained by subtracting a constant amount from the volume. This corresponds to the volume of solids, plus a small part of the water and is found to be about 30 μm^3. Thus,

$$(\text{volume in } \mu m^3 - 30) \times \%\,\text{NaCl} = \text{Constant.} \qquad (3.2)$$

3.1.6 **The above erythrocyte has a volume of 84 μm^3 in 0.9% NaCl. At what concentration would it swell to a sphere of 156 μm^3 and haemolyse?**

Is this a realistic answer? Recall that the concentration at which 50% of corpuscles haemolyse is typically about 0.37–0.44% NaCl.

In the final stage of haemopoeisis the nuclei of normoblasts become more compact (pyknotic) and are lost. How would the corpuscular volume differ if the nucleus were retained (as in reptiles, birds and other non-mammalian vertebrates)? Interphase nuclei vary in size from one kind of cell to another, but for present purposes we can choose to postulate any reasonable volume for what is just a hypothetical erythrocyte nucleus. Let us therefore start from the familiar. A small lymphocyte, as seen in a blood film, consists mostly of nucleus and has a diameter close to that of the erythrocyte. Ignoring the fact that leucocytes tend to spread out a little on the slide, but also may shrink a little in preparation, let us take the nuclear diameter as 7 μm. (Erythrocytes and nuclei are usually to be seen together in histological sections; are they generally of roughly similar diameter?) The volume of a sphere of diameter 7 μm is $(7^3 \times \pi/6)$, i.e. 180 μm^3.

We may now imagine inserting a nucleus of this volume into the erythrocyte of Table 3.1, this having initially a volume of 84 μm^3.

3.1.7 **What would be the final volume of this erythrocyte?**

If this nucleated erythrocyte were spherical, it would have a diameter of 8 μm.

3.1.8 **For a given haematocrit, by what factor would the haemoglobin content be reduced by the presence of the nuclei?**

3.2 Optimum haematocrit – the viscosity of blood

The more erythrocytes there are in a given volume of blood, the more oxygen can that blood contain. On the other hand, the viscosity of the blood increases with increasing haematocrit and the blood flows less easily (equation 3.1). One might guess, therefore, that there exists an optimum haematocrit for oxygen carriage.

Viscosity does not increase linearly with haematocrit, but rises more and more steeply as haematocrit increases. Thus, in a particular set of experiments, for haematocrits of 20, 35, 52 and 62%, corresponding viscosities were found to be 1.4, 2.0, 3.0 and 4.0 centipoise (Fig. 3.3). It does not matter if the units are unfamiliar. Since flow rates vary inversely with viscosities, the relative rates at which erythrocytes (and their contained oxygen) flow, e.g. through a blood vessel, may be found by dividing the four haematocrit values by the respective viscosities.

3.2.1 **What are these ratios for the four haematocrit values (20, 35, 52 and 62%)?**

It would seem from these calculations that the optimum concentration of erythrocytes in the blood corresponds to the normal range of values (as given in Section 3.1) – a pleasing result! As so often in physiology,

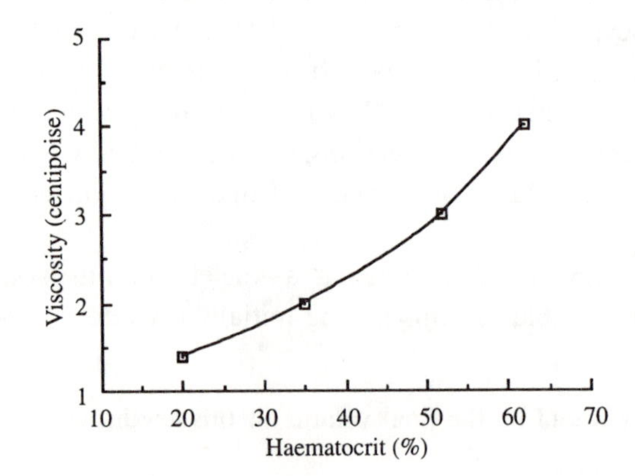

Fig. 3.3. The data of question 3.2.1 – showing how blood viscosity (in centipoise) rises with haematocrit.

the matter is not as simple as it might appear. A given sample of blood does not actually have a fixed viscosity that is independent of how and where it is flowing and measurements of blood viscosity therefore depend on the kind of viscometer that is used. If, for example, one times the flow of blood through a capillary tube, then the estimated viscosity depends on the tube's internal diameter. Fortunately, the method of measurement makes less difference to the estimated optimum haematocrit; the haematocrit at which the ratio of haematocrit to viscosity is maximum works out typically as about 40–50%.

No account has been taken of the fact that oxygen is carried by the blood in simple solution as well as in combination with haemoglobin. Strictly, the above calculations should have taken account of that. However, of the typical 200 ml oxygen/l of arterial blood, only 3 ml/l is in simple solution and the correction is not worth making.

3.3 Peripheral resistance

The rate at which blood flows through any part of the cardiovascular system increases with the pressure difference between the two ends and decreases with the resistance to flow. As noted above,

$$\text{flow} = \text{pressure difference}/\text{resistance}. \qquad (3.1)$$

This commonly stated relationship neglects the kinetic energy of the blood and also its gravitational potential energy (Chapter 1), but suffices for the present purpose. In this Section the equation is applied to the whole systemic and pulmonary circulations. Blood pressures are assumed to be measured at heart level, so that differences in gravitational potential energy do not need to be considered.

Equation (3.1) can be re-written as:

$$\text{cardiac output} = \frac{\text{mean aortic pressure} - \text{mean venous pressure}}{\text{Resistance}}. \qquad (3.3)$$

In the case of the systemic circulation it is customary to regard the central venous pressure as negligibly small compared with the arterial pressure so that, by rearrangement of equation (3.3),

$$\text{mean aortic pressure} = \text{cardiac output} \times \text{peripheral resistance}. \qquad (3.4)$$

This important relationship highlights cardiac output and peripheral resistance as determinants of blood pressure. Less often given in elementary accounts are any actual values for the peripheral resistance. This is because they are not particularly memorable in themselves and have unfamiliar units (e.g. mmHg·min/l). Peripheral resistance is in any case immediately calculable from the other, better-known, quantities. Thus, if the mean aortic blood pressure at rest is taken as 100 mmHg and the cardiac output is 5 l/min, then the peripheral resistance is:

$$\frac{100\ \text{mmHg}}{5\ \text{l/min}} = 20\ \text{mmHg}\cdot\text{min/l}.$$

The resistance is sometimes given in '*R* units', equivalent to mmHg·sec/ml. The above value in *R* units is $20 \times 60/1000 = 1.2$. With a cardiac output of 6 l/min, the resistance is a convenient 1.0 *R* unit.

The above values of peripheral resistance are more obviously meaningful when compared with others, such as those that apply in exercise and in the pulmonary circulation. If we deal with relative resistances as before, we may ignore the units.

3.3.1 **If the cardiac output rises threefold during exercise, with mean aortic blood pressure rising from 100 mmHg to 108 mmHg, by what factor does the peripheral resistance fall?** (Is the change in pressure worth including in the circulation?)

For the pulmonary circuit, the venous pressure is too close to the arterial pressure to be neglected. It is therefore necessary to use equation (3.3).

3.3.2 **Taking the mean pressures in the pulmonary arteries and veins at rest as 12 mmHg and 5 mmHg, respectively, what is the pulmonary resistance as a percentage of the systemic peripheral resistance given above?**

In exercise, the pulmonary arterial pressure may rise, stay constant, or even fall slightly, depending on the nature and timing of the exercise. Changes in pulmonary vascular resistance are also variable.

3.4 Blood flow and gas exchange

3.4.1 **For a blood volume of 5 l and a cardiac output of 5 l/min, what is the circulation time, that is to say, the average time required for a blood cell to pass completely around the systemic and pulmonary circuits?**

This answer is appropriate to a resting human body. The corresponding time in a shrew (with a heart rate of about 10 beats/s) is only about 3 s – reducing to about 1 s in exercise.

For a given cardiac output the average velocity of blood in a particular category of blood vessel (aorta, arteriole, etc.) is inversely proportional to the total cross-sectional area. The cross-sectional area of the aorta is about 4.5 cm^2 and the total cross-sectional area of the capillaries (variable and impossible to measure directly) has been estimated as 4500 cm^2.

3.4.2 **Accordingly, what is the average velocity of the blood in cm/s, in (a) the aorta and (b) the capillaries, when the cardiac output is 90 ml/s (5.4 l/min)?**

3.4.3 **If a typical capillary length is taken as 0.5 mm, how long does it usually take an erythrocyte to pass through?** (Ignore axial streaming, the tendency of corpuscles to travel down the middle of a vessel at a higher velocity than plasma near the endothelium; in capillaries there is often no plasma between erythrocyte and endothelium.)

That answer is for a non-exercising person; in exercise, the time spent by each erythrocyte in a capillary is very much shorter. The time is also likely to be shorter in a smaller mammal because of the higher cardiac output of smaller animals (Section 2.7). That the blood velocity within capillaries is at least close to being a limiting factor in the processes of gas exchange is suggested by several facts. Thus the haemoglobin of smaller mammals releases oxygen more readily at a given pH and the Bohr effect is bigger. Carbonic anhydrase activity is also greater in the blood of small mammals compared with large ones, allowing more rapid exchange of carbon dioxide.

Cardiac output may be estimated, using the Fick principle, from the measured rate of oxygen consumption and the oxygen contents of the arterial and mixed venous blood. Thus,

$$\text{cardiac output} = \frac{O_2 \text{ consumption}}{\text{arterial } O_2 - \text{mixed venous } O_2}. \tag{3.5}$$

The same formula may be turned around to calculate the arterio-venous difference in oxygen content from cardiac output and rate of oxygen consumption. If all the quantities in the next question and answer are familiar, they may yet be worth checking for compatibility.

3.4.4 **If the cardiac output of a man is 5 l/min, if his oxygen consumption is 0.25 l/min (Section 2.5), and if the oxygen content of his arterial blood is a typical 200 ml/l (Section 2.4), what is the oxygen content of his mixed venous blood?**

The delivery of oxygen to the tissues in exercise is accomplished partly by an increase in cardiac output and partly by an increase in the amount of oxygen extracted from the blood. Suppose that the man described in question 3.4.4 increases his oxygen consumption from 0.25 to 2.8 l/min, that his cardiac output increases, not proportionately from 5 to 56 l/min, but only to 20 l/min (these being representative, non-extreme changes in strenuous exercise) and that the oxygen content of his arterial blood stays at 200 ml/l.

3.4.5 **What is now the oxygen content of his mixed venous blood?**

3.5 Arteriolar smooth muscle – the Law of Laplace

The Law of Laplace, as applied to a cylindrical tube, relates the circumferential wall tension per unit length (T) to the difference between the internal and external pressures (P) and the radius (r):

$$P = T/r. \tag{3.6}$$

Here we assume the external pressure to be zero, so that P is simply the internal pressure, i.e. the blood pressure in the case of a blood vessel.

This formula is often quoted in order to explain why a capillary with its thin and flimsy wall does not burst; if a capillary has a radius of 4 μm and an artery has a radius of 4 mm, the wall tension for a given blood pressure, or T/P, is 1000-fold lower in the capillary than in the artery. Less often are actual tensions calculated and units therefore specified; as expressed above, without any added numerical factor, the units are SI. Thus, with T in units of N/m and r in m, P is in N/m^2.

Quantification is generally useful only when there is a comparison to be made and here we relate the tension in the wall of an arteriole due to blood pressure and the opposing tension that can be generated by its smooth muscle. In doing this we actually deal here, not with tensions or stresses as such, but with the mean circumferential stress per unit area of wall thickness, S. For a wall thickness w, T is wS and:

$$P = wS/r. \tag{3.7}$$

Blood pressures (P) in N/m^2 can be converted to the more familiar units of mmHg by multiplying by $7.5(02) \times 10^{-3}$. The units N/m^2 can be converted to kg/cm^2, by multiplying by $1(02) \times 10^{-5}$. Units of kg/cm^2 may be the more familiar ones in the context of muscle tensions.

Here is equation (3.7) re-expressed so that P is in mmHg and S is in kg/cm^2; w and r can be in any units so long as these are the same for both.

$$P \text{ (in mmHg)} = 735 \times w/r \times S \text{ (in kg/cm}^2\text{)}. \qquad (3.8)$$

The value 735 has units of $mmHg \cdot cm^2/kg$ and can be derived from Table 1.1.

We may now consider a small arteriole in which the smooth muscle layer is only one cell thick and estimate the maximum internal pressure against which the smooth muscle can contract. For simplicity we may, for the moment, think of the arteriole wall as being made up only of the smooth muscle cells (disregarding the thickness of endothelium and connective tissue and the elasticity of the elastin). Let us take a radius, r, of 15 μm and a wall (muscle cell) thickness, w, of 5 μm, and assume that the muscle cells can develop a tension of 3 kg/cm^2. Of course, all three of these vary during vasodilatation and vasoconstriction, as well as from vessel to vessel. In particular, the length–tension curve for vascular smooth muscle shows a maximum tension (of perhaps 3.5 kg/cm^2), with decreasing tensions at lengths that are longer and shorter than optimum.

3.5.1 **On the basis of these values, what is the pressure, *P*, within the arteriole that is required to balance the tension in the active smooth muscle?**

Actual blood pressures in the smallest arterioles vary, but are typically about 20–50 mmHg at heart level. Even allowing for an extra 100 mmHg in the feet of a standing person, it seems that the single layer of smooth muscle cells should be able to cope! This conclusion is clearly not very sensitive to the initial assumptions and simplification, but what if the vessel were much more dilated? Dilatation involves an increase in r and decrease in w, so that the value of w/r is reduced.

3.5.2 **By what factor could the ratio of wall thickness to vessel radius, w/r, decrease before contraction becomes impossible against an internal pressure of 150 mmHg, if the muscle cells still produce a tension of 3 kg/cm^3?**

3.6 Extending William Harvey's argument

'What goes in must come out'

William Harvey calculated that the heart pumps out so much blood in half an hour that only circulation makes sense (see the Preface). We have, unlike fishes, a 'double circulation', with separate pulmonary and systemic circuits. In the long term, the flow through the two circuits must be equal.

3.6.1 **A man has a blood volume of 5.00 l. Momentarily, the output from his right ventricle is 5.00 l/min and his left ventricular output is 4.95 l/min. If this small (1%) discrepancy could persist for a while, and the blood volume were to stay constant, how much extra blood would the pulmonary vessels contain after 20 min?**

The answer may be compared with the half litre of blood that would normally be present in his pulmonary blood vessel. Though the situation is unreal and the conclusion perhaps obvious, this calculation does at least highlight the need for some control mechanism to keep in long-term balance the output from each side of the heart. This is the Frank–Starling mechanism whereby the stroke volume of each ventricle increases with diastolic filling, and hence with venous return (Starling's Law of the Heart). The same kind of argument may be applied wherever a balance of input and output must be maintained: body water balance, the balance of carbon dioxide production and loss, heat balance, the matching of transmembrane transport processes at the two sides of an epithelial cell (as shown in Fig. 5.4), and so on.

3.7 The work of the heart

With the body at rest, the oxygen consumption of the heart is commonly said to be about 8–10 ml/min per 100 g. For a heart of 300 g it would thus be about 24–30 ml/min. (This may be compared with the basal oxygen consumption of the whole body, which is usually near 250 ml/min.) With the myocardium using as fuel about 40% carbohydrate and 60% fatty acids, 1 ml of oxygen is equivalent to about 4.8 cal (20 J). An oxygen consumption of 24–30 ml/min is thus equivalent to 115–144 cal/min (480–600 J/min). These last values may be compared with estimates of the rate at which the heart does external work (its power output). For each side of the heart this is calculable from the following equation, in which IBP stands for the increase in mean dynamic pressure as the blood passes through the heart:

$$\text{power output} = \text{cardiac output} \times \text{IBP}. \tag{3.9}$$

An expression for the dynamic pressure of flowing fluid is given in Chapter 1, i.e. $(P + \rho gh + \frac{1}{2}\rho v^2)$. Here we can simplify this expression by ignoring the work done against gravity (hence the term ρgh) and also the kinetic energy term ($\frac{1}{2}\rho v^2$). That the kinetic energy of the blood is indeed negligible in the healthy, non-exercising body that is our present concern is a point we return to later. With these simplifications, we need to consider only the increase in mean blood pressure produced by the heart, that is to say the difference between the mean arterial and venous pressures. Furthermore, when the venous pressure is close to zero, as it usually is, equation (3.9) may be simplified even further to:

$$\text{power output} = \text{cardiac output} \times \text{mean arterial pressure}. \tag{3.10}$$

In terms first of SI units, and then of other units (Table 1.1), this can be represented as

$$\frac{\text{J}}{\text{s}}\ (\text{i.e. watts}) = \frac{\text{N m}}{\text{s}} = \frac{\text{m}^3}{\text{s}} \times \frac{\text{N}}{\text{m}^2}$$

or

$$\text{J/min} = \frac{\text{l/min}}{1000} \times (\text{mmHg} \times 133.3)$$

$$= (\text{l/min}) \times (\text{mmHg}) \times 0.13,$$

or, in terms of calories,

$$\text{cal/min} = (\text{l/min}) \times (\text{mmHg}) \times 0.032.$$

3.7.1 **On the basis of these relationships, how much external work does the left heart do per minute (in cal/min or J/min) when the cardiac output is 5 l/min and the mean aortic blood pressure is 100 mmHg)**

As to the right side of the heart, the volume output is the same, but the pressures are much lower. Blood enters the right atrium at a pressure not far from zero and leaves the right ventricle at a mean pulmonary arterial pressure of about 12–15 mmHg. The external power output of the right heart is thus only 12–15% of that of the left heart.

The total of the two values of power output (say 18 cal/min, 73 J/min) may be compared with the values given above for the actual energy expenditure of the heart, namely 115–144 cal/min (480–600 J/min). The disparity may be expressed in terms of the percentage 'efficiency' of pumping, calculated as

$$\frac{\text{external work/min}}{\text{total work/min}} \times 100.$$

3.7.2 **If the total external work done per minute by the two sides of the heart is 18 cal/min (73 J/min) and the actual energy expenditure is 130 cal/min (534 J/min), i.e. within the range just given, what is the percentage efficiency?**

The efficiency thus calculated is fairly typical for basal conditions and the obvious point to note is that it is quite low. It does, however, increase substantially when the heart works harder, as in exercise. It may be lower under pathological conditions. The variable efficiency implies that there is a poor correlation between external and total work.

We now return to the kinetic energy term hidden in equation (3.9) that was taken as zero. This is $\frac{1}{2}\rho v^2$, where ρ is the density of the blood and v is blood velocity. Here v^2, and not v, needs to be averaged over systolic and diastolic velocities. Because $\frac{1}{2}\rho v^2$ increases with the square of velocity, the term can become very significant in equation (3.9) when the cardiac output is increased in exercise, and also in aortic stenosis. However, our concern here is to show that the term is normally small at resting levels of cardiac output. To this end, we may start by ignoring the complications of pulsatile flow and utilize the square of the mean of v rather than the mean of v^2. The fact that the units may not be obviously compatible with those for P, namely N/m^2 (or Pa) is discussed in Chapter 1; with ρ in units of kg/m^3, and v in m/s, $\frac{1}{2}\rho v^2$ does indeed work out in units of N/m^2. Divided by 133.3, this becomes mmHg.

The mean velocity of the blood in the aorta was calculated in question 3.4.2 as 20 cm/s (0.2 m/s). The density of blood is about 1050–1064 kg/m^3, but may be taken here as 1000 kg/m^3.

3.7.3 **What is $\frac{1}{2}\rho v^2$ in mmHg when v is 0.2 m/s?**

The point is made; the kinetic energy term is indeed very low compared with the mean blood pressure of about 100 mmHg. However, because the blood flow is actually pulsatile, the calculation is valid only insofar as it indicates the order of magnitude of $\frac{1}{2}\rho v^2$. The answer is actually too low.

It has already been pointed out that the calculation should utilize, not the mean velocity, but the mean of its square, and this is less easily determined. What follows is not an attempt to obtain an exact answer, but rather an illustration of the difference between the two approaches. It is intended less as sound physiology and more as the making of an arithmetical point.

Assume that the mean velocity is again 0.2 m/s, but that flow is, not actually pulsatile, but interrupted; for one-third of the time the velocity is $3 \times 0.2 = 0.6$ m/s and for two-thirds of the time the velocity is zero.

3.7.4 **What is the mean value of v^2?**

For comparison, recall that for steady flow at the same mean velocity, v^2 is 0.04 m^2/s^2. With the flow regarded as intermittent, the value of $\frac{1}{2}\rho v^2$ is three times as great. Note, however, that it is still small enough to be neglected for the purposes above.

4

Respiration

Sections 4.1 to 4.5 are about oxygen and carbon dioxide in blood, cells and alveoli. Section 4.1 is a reminder of how to correct gas volumes for variations in temperature, pressure, etc., but is more about how and when to avoid such calculations. From a straightforward treatment of how concentrations of dissolved gases relate to partial pressures, we move on to consider the significance of cytoplasmic carbon dioxide tensions (Section 4.3), alveolar gas tensions at altitude (Section 4.4) and why the alveolar arterial carbon dioxide tensions are so much higher in mammals than in fish (Section 4.5). The latter topic relates to that of Section 4.6 – the loss of water in expired air. Sections 4.7 to 4.9 are about breathing and the structure and dimensions of the lungs. Sections 4.10 to 4.12 are concerned with surface tensions and fluid pressures in the lungs and pleural space, and hence with pulmonary oedema.

On the mathematical front, the calculations continue to use little more than shop-keeper's arithmetic. However, geometric progressions appear here in the two contexts of airway branching (Section 4.9) and the renewal of alveolar gas (Section 4.7) and the latter topic provides an easy introduction to exponential time courses (for which see also the subject of renal drug clearance discussed in Section 5.3).

4.1 When not to correct gas volumes for temperature, pressure, humidity and respiratory exchange ratio

In analysing experimental results it is necessary to correct for the effects on gas volumes of temperature, pressure and humidity. Three sets of conditions are of particular interest:

ATPS: ambient temperature and pressure, saturation with water vapour at ambient temperature.

BTPS: body temperature, alveolar gas pressure (taken as ambient), saturation with water vapour at body temperature.

STPD: standard temperature (0 °C, 273 K) and pressure (760 mmHg), dry.

In the context of the often rough-and-ready calculations of this book, these corrections may be just a hindrance; after all, the raw data to be used tend to be 'typical values' or round numbers. The formulae for making the corrections are given below for reference, but the main point of introducing them is to explore the general magnitude of the corrections so that one may judge their importance. Nothing emerges in this Section concerning body function.

The corrections are based on the following relation, in which P, V and T are respectively the pressure, volume and absolute temperature of the gas or gas mixture, n is the number of moles of gas and R is the gas constant:

$$PV = nRT. \tag{4.1}$$

The calculations can be complicated by the fact that the value of n, which refers to water vapour as well as the other gases in a mixture, does not stay constant. However, considering only the dry gas, we have, for two conditions,

$$P_1V_1/T_1 = P_2V_2/T_2. \tag{4.2}$$

This is also valid for moist gases provided that water vapour pressures are subtracted from P_1 and P_2. Thus, V_{BTPS} can be converted to V_{STPD} from the following formula, in which body temperature is taken as 37 °C and the corresponding water vapour pressure is 47 mmHg. The total pressure in the alveoli is the same as the ambient (barometric) pressure, P_B.

$$\frac{V_{STPD}}{V_{BTPS}} = \frac{[P_B - 47\ \text{mmHg}]\cdot[273]}{[760\ \text{mmHg}]\cdot[273 + 37]}$$

$$= [P_B - 47\ \text{mmHg}] \times 6.16 \times 10^{-4}. \tag{4.3}$$

The formula for the interconversion of V_{BTPS} and V_{ATPS} may be derived from equation (4.2) similarly and is as follows, T_{amb} being the ambient temperature in degrees Celsius and $P_{amb(H_2O)}$ the ambient water vapour pressure.

$$\frac{V_{ATPS}}{V_{BTPS}} = \frac{[P_B - 47\ \text{mmHg}]\cdot[273 + T_{amb}]}{[P_B - P_{amb(H_2O)}]\cdot[273 + 37]}. \tag{4.4}$$

So long as the body is warmer than the environment, the ratio of V_{ATPS} to V_{BTPS} is less than 1.0. The ratio increases with the ambient temperature.

4.1.1 **If the ambient temperature is 10 °C, at which the water vapour pressure at saturation is 9 mmHg, and the barometric pressure, P_B, is 750 mmHg, what is the ratio V_{ATPS}/V_{BTPS}?**

4.1.2 **At that barometric pressure and ambient temperature, the latter deliberately rather low for a laboratory, what is the percentage error if V_{BTPS} is incorrectly taken as being the same as V_{ATPS}?**

Clearly, the correction does matter for accurate work.

Equation (4.4) includes three quantities relating to the environment, namely P_B, T_{amb} and $P_{amb(H_2O)}$. If the barometric pressure, P_B, in 4.1.1 is altered to any value between 634 and 918 mmHg, the answer works out the same to within 1%. In other words, P_B does not have to be accurately know under normal conditions near sea level. On the other hand, variations in T_{amb} make considerable difference. The effect of this term in equation (4.4) may be seen by comparing, say, 0 and 40 °C: the ratio of (273 + 40) to (273 + 0) is 1.15.

Temperature also affects the value of $P_{amb(H_2O)}$. What if one does not know the correct water vapour pressure, but knows only that it must be somewhere between zero and the value for 37 °C (namely 47 mmHg)? Suppose that one simply takes a mid-way value of 47/2 mmHg, so that, with P_B equal to 760 mmHg, $[P_B - P_{amb(H_2O)}]$ is 736.5 mmHg. The maximum percentage error in V_{ATPS}/V_{BTPS} is then equal to $(760/736.5 - 1) \times 100$ or $(1 - 713/736.5) \times 100$, these being the same.

4.1.3 **What is the maximum percentage error in V_{ATPS}/V_{BTPS} if the water vapour pressure is taken as 23.5 mmHg?**

Equation (4.4) implies that a given mass of air increases in volume as it is inspired, through both warming and moistening. A greater volume is expired than is inspired. The effect is no longer apparent once both volumes are expressed in the same terms (BTPS, ATPS or STPD). However, a small discrepancy may remain, due to the fact that less carbon dioxide is given out, typically, than oxygen is extracted. The ratio of carbon dioxide given out from the body to oxygen extracted from the air is the respiratory exchange ratio (**R**). In the steady state, this is the same as the whole-body respiratory quotient (RQ), and usually near 0.8. How important is this effect?

As an example, suppose that the dried inspired air contains 21% oxygen, 79% nitrogen, and virtually no carbon dioxide, and that dried expired air contains 16% oxygen and 4% carbon dioxide (corresponding to **R** of about 0.8, but see below), hence 80% nitrogen. Given that the combined percentage of oxygen and carbon dioxide falls from (21 + 0)% to (16 + 4)%, it would seem that the reduction in total volume is about 1%. In other words, it is for many purposes too small to matter. That is the main point to be made here and the reader may be content with that. However, the values of both **R** and the change in volume have still to be calculated properly.

For that purpose, we assume that as much nitrogen is breathed in as is breathed out and that all volumes are measured dry and at one temperature. In 100 l of inspired air there are 21 l of oxygen and 79 l of nitrogen. When exhaled, this 79 l of nitrogen constitutes 80% of the expired air. It is accompanied by $(16\%)/(80\%) \times 79 = 15.80$ l of oxygen and $(4\%)/(80\%) \times 79 = 3.95$ l of carbon dioxide.

4.1.4 **Considering dry gases at one temperature, what is the expired volume as a percentage of the inspired 100 l?**

4.1.5 **What is the respiratory exchange ratio, R?**

4.2 Dissolved O_2 and CO_2 in blood plasma

The concentration of a gas in simple solution increases with the partial pressure (tension) of the gas in question (e.g. P_{CO_2}, P_{O_2}) and with the solubility coefficient (e.g. S_{CO_2}, S_{O_2}). Thus, using square brackets to denote concentrations,

$$[\text{dissolved } CO_2] = S_{CO_2} \cdot P_{CO_2} \tag{4.5}$$

and

$$[\text{dissolved } O_2] = S_{O_2} \cdot P_{O_2}. \tag{4.6}$$

For a discussion of units see Chapter 1.

The following questions are to ensure familiarity both with these relationships and with normal arterial gas tensions, all of which are used in later sections. S_{CO_2} and S_{O_2} decrease with increasing temperature, but the questions utilize values appropriate to human plasma at 37 °C.

4.2.1 Given a normal arterial P_{CO_2} of 40 mmHg and S_{CO_2} of 0.03 mmol/l per mmHg, what is the concentration of dissolved CO_2 in plasma in mmol/l?

Note how low this value is compared with the concentration of HCO_3 in the plasma. That is about 25 mmol/l.

4.2.2 Given a normal arterial P_{O_2} of 100 mmHg and S_{O_2} of 0.0014 mmol/l per mmHg, what is the concentration of dissolved O_2 in plasma in mmol/l?

The answer is about the same for whole blood as for plasma. It may be compared with the total concentration of O_2 in fully oxygenated whole blood of about 9 mmol/l (equivalent to 200 ml/l) – see 2.4.1).

4.3 P_{CO_2} inside cells

The value of P_{CO_2} is generally 40 mmHg in arterial blood and about 46 mmHg in mixed venous blood. What might it be inside a 'typical' cell? One approach to this question is to consider the gradient of carbon dioxide between blood and cells and to relate it to that of oxygen. Let us assume that the two gases move in opposite directions along the same path between cell and blood and do so only by simple diffusion (Fig. 4.1). The rate of diffusion for each (along any given path) is proportional to the respective difference in partial pressure multiplied by the appropriate diffusion coefficient (see Chapter 1 for more on such coefficients). Here we do not need to know actual diffusion coefficients – only that the value for carbon dioxide that is appropriate to cells and extracellular fluid is about 20-fold that for oxygen.

In a steady state the rates of diffusion must also equal the rates of utilization (for oxygen) and production (for carbon dioxide). We need only to know relative, not actual, rates. The rate of carbon dioxide production divided by the rate of oxygen consumption is the respiratory quotient (RQ).

All of this combines to yield the following equation, in which the subscripts b and c refer to blood and cells respectively:

$$(P_{cCO_2} - P_{bCO_2}) = (P_{bO_2} - P_{cO_2}) \times RQ/20. \qquad (4.7)$$

This may seem unhelpful at first if P_{cO_2} is also unknown, but we can at

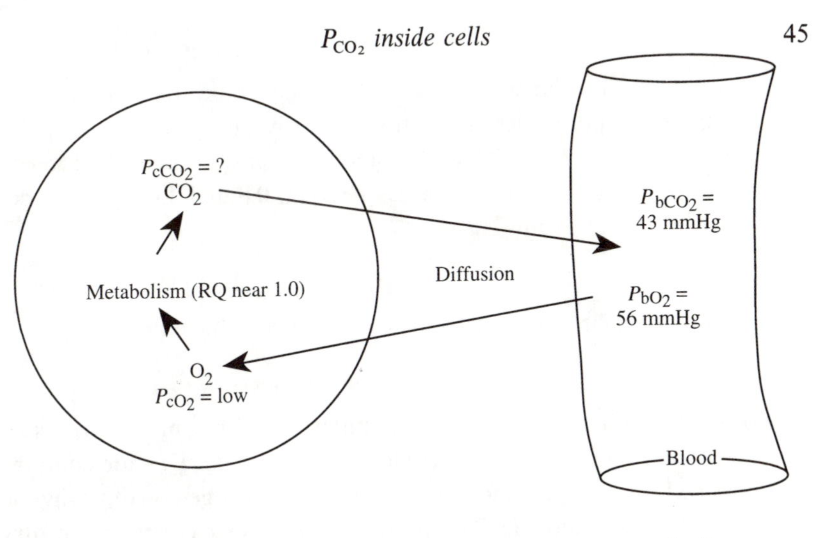

Fig. 4.1. Oxygen and carbon dioxide diffusion gradients between a cell and a nearby capillary: the gas tensions in the blood are as assumed in questions 4.3.1 and 4.3.2.

least try setting a lower limit to the latter, namely zero. For the gas tensions in blood let us take values between arterial and mixed venous tensions. A comparison of oxygen and carbon dioxide dissociation curves for blood in any textbook suggests that a suitable combination of P_{bCO_2} and P_{bO_2} would be 43 and 56 mmHg, respectively. (As will become apparent, the exact values matter little.)

4.3.1 **What is the value of ($P_{cCO_2} - P_{bCO_2}$) when $P_{cO_2} = 0$, $P_{bO_2} = 56$ mmHg and RQ = 1?**

4.3.2 **Taking P_{bCO_2} as 43 mmHg, what is the value of P_{cCO_2}?**

The answer might be too high to be taken as typical since the RQ is somewhat high. Moreover, if P_{cO_2}, at zero, must be too low. Estimates exist for P_{cO_2} in particular tissues, for example about 2–3 mmHg in sarcoplasm of red skeletal muscle, but the whole-body average cannot be measured. Let us take a credible, non-zero value for P_{cO_2} of 10 mmHg.

4.3.3 **Accordingly, what is the value of P_{cCO_2} when RQ = 0.8, $P_{cO_2} = 10$ mmHg, $P_{bO_2} = 56$ mmHg and $P_{bCO_2} = 43$ mmHg?**

Compare this with the answer to 4.3.2. One could try out various combinations of input values in equation (4.7), without obtaining markedly different answers for P_{cCO_2}. Thus, for blood gas tensions between arterial and mixed venous, and for RQ values between 0.8 and 1.0, P_{cCO_2} works out between 41 and 51 mmHg.

4.4 Alveolar gas tensions at sea level and at high altitude

Inspired air (standard values)

Standard atmospheric pressure is 760 mmHg (101.3 kN/m^2 or 101.3 kPa – near enough 100 for back-of-envelope calculations). Dry air contains just under 21% of oxygen, most of the rest being nitrogen, with negligible amounts of carbon dioxide. The partial pressure of oxygen, P_{O_2}, in dry air at standard pressure is thus 21% of 760 mmHg, i.e. 160 mmHg.

Inspired dry air becomes warmed and saturated with water vapour in the airways. The water vapour pressure at 37 °C (i.e. the partial pressure of water vapour) is 47 mmHg, so that the partial pressure of all other gases taken together is (760 – 47) = 713 mmHg. The partial pressure of oxygen is 21% of that, i.e. 150 mmHg.

Alveolar gas near sea level

As already noted, the respiratory exchange ratio, **R**, is the ratio of carbon dioxide output from the lungs to oxygen uptake. It is related to the partial pressures of carbon dioxide and oxygen in alveolar gas, P_{ACO_2} and P_{AO_2}, and the partial pressures of carbon dioxide and oxygen in the warmed and moistened inspired air, P_{ICO_2}, and P_{IO_2}, as follows:

$$\mathbf{R} = (P_{ACO_2} - P_{ICO_2})/(P_{IO_2} - P_{AO_2}). \quad (4.8)$$

(Strictly, a small correction is needed when the volumes of inspired and expired gases are not exactly equal.) By rearrangement, and taking the values P_{IO_2} as 150 mmHg and P_{ICO_2} as zero,

$$P_{ACO_2} = \mathbf{R}(150 - P_{AO_2}). \quad (4.9)$$

4.4.1 **Is this equation compatible with textbook values of P_{AO_2} and P_{ACO_2}? Specifically, what is the value of R if P_{AO_2} and P_{ACO_2} are, respectively, 100 and 40 mmHg?**

For a carbohydrate diet, **R** is close to 1 and for a diet mainly of fat *R* is close to 0.7 (Table 2.1).

Expired air near sea level

The mean composition of expired air varies with the depth of breathing and is therefore less memorable. However, the approximate value of P_{CO_2} needs to be known for Section 4.5. The expired air is a mixture of alveolar gas and dead space air (which, at the start of expiration, is warmed and moistened inspired air with almost zero P_{CO_2}).

4.4.2 **For a (typical) tidal volume of 500 ml and dead space of 150 ml, what is the mean P_{CO_2} in the expired air if the alveolar P_{CO_2} is 40 mmHg?**

More usefully, the calculation may be reversed – to calculate the volume of the dead space.

Altitude

Air pressure decreases with increasing altitude. Some South Americans live at a height of about 18 000 feet (3.4 miles or 5 500 m) above sea level where the barometric pressure is about 380 mmHg. In air moistened at 37 °C, the partial pressure of water vapour is again 47 mmHg, so that the combined pressure of other gases is 333 mmHg. Of these, oxygen still constitutes about 21%.

4.4.3 **What is the partial pressure of oxygen in moist inspired air at 18 000 feet?**

The low P_{O_2} in the inspired air results in a low P_{O_2} in the alveoli also. Equation (4.9) applies again, except that the value of 150 mmHg (i.e. P_{IO_2}) must be replaced by the answer to 4.4.3.

4.4.4 **At the same altitude, if alveolar P_{CO_2} were to be maintained at 40 mmHg and R were 0.9, what would alveolar P_{O_2} be?**

The critical alveolar P_{O_2} at which the average individual loses consciousness on brief exposure to hypoxia is about 30 mmHg. However, the hypoxia leads to an immediate increase in pulmonary ventilation and so a fall in alveolar P_{CO_2}. P_{O_2} would therefore rise correspondingly, to a value somewhat above that just calculated. After acclimatization, the P_{CO_2} would be below 30 mmHg instead of 40 mmHg.

4.5 Why are alveolar and arterial P_{CO_2} close to 40 mmHg?

One kind of answer to this question is that P_{CO_2} is kept near that value by homeostatic mechanisms – but why was that particular value favoured during our evolution? Arterial P_{CO_2} must have been very much lower in our aquatic ancestors, as it is in present-day fish. Here we consider first why P_{CO_2} has to be low in water-breathing animals. We then look at one of the implications of having a higher arterial P_{CO_2}.

Think of a fish living in well-aerated, carbon-dioxide-free water containing oxygen at a tension of 160 mmHg, as in the air we breathe. Assume, for simplicity, that the fish releases carbon dioxide at the same rate as it uses oxygen (i.e. $\mathbf{R} = 1$, defined as in Sections 4.1 and 4.4). Then, as the water passes the gills, the change in its carbon dioxide content is equal and opposite to the change in oxygen content. In accordance with equations (4.5) and (4.6), each of these changes is equal to the change in partial pressure multiplied by the solubility of the gas in question (i.e. S_{O_2} or S_{CO_2}). Therefore, the partial pressures of carbon dioxide and of oxygen in the water leaving the gills are related thus:

$$S_{CO_2} \times (P_{CO_2} - 0) = S_{O_2} \times (160 - P_{O_2}), \tag{4.10}$$

or, rearranging,

$$P_{CO_2} = S_{O_2}/S_{CO_2} \times (160 - P_{O_2}). \tag{4.11}$$

Carbon dioxide is much more soluble in water than is oxygen. Indeed, S_{O_2}/S_{CO_2} in the fish's aquatic environment would be between about 1/25 (at 30 °C) and 1/35 (at 0 °C).

4.5.1 **What would be the maximum P_{CO_2} in the water leaving the gills, corresponding to complete extraction of oxygen?**

The P_{CO_2} of the blood leaving the gills could, in principle, be higher than that, as a result of incomplete equilibration or through the admixture of 'venous' blood, but it would certainly have to be much lower than 40 mmHg.

Equation (4.10) may be adapted to the mammalian (air-breathing) condition by setting S_{O_2}/S_{CO_2} equal to 1.0 (because the two gases are equally 'soluble' in air) and by taking the oxygen tension of the respiratory medium (i.e. warm, moist air) as 150 mmHg instead of 160 mmHg. At this point it is useful to include a term that was taken as unity in equation (4.10) and therefore omitted, namely the respiratory exchange ratio, **R.** We then have:

$$P_{CO_2} = \mathbf{R} \times (150 - P_{O_2}). \tag{4.12}$$

Applied to alveolar gas, equation (4.12) is the same as equation (4.9).

4.5.2 **What is the limiting (maximum) value of P_{ACO_2} for R = 0.8, namely P_{ACO_2} when alveolar P_{ACO_2} is zero?**

The equation thus tells us that P_{CO_2} can be much higher in mammals than in fishes, but that there is some ill-defined upper limit set by the minimum tolerable P_{AO_2}. As noted in Section 4.4, the minimum tolerable P_{AO_2} exceeds 30 mmHg (corresponding to P_{ACO_2} = 96 mmHg), but it is obviously less than 100 mmHg. Equation (4.12) does not allow exact calculation of our alveolar and arterial carbon dioxide tensions.

One reason why we have not retained the low carbon dioxide tensions of our water-breathing ancestors is probably that any surface for respiratory exchange in an air-breathing animal is necessarily also a route for evaporative water loss. Any anatomical arrangement, such as an enclosed lung, that reduces evaporation also results in higher internal tensions of carbon dioxide. The evaporation of water implies also the loss of heat (the latent heat of evaporation, L, in kcal/mol or kJ/mol) and this is particularly important for homeotherms like ourselves. Consider the case of a mammal that is inhaling air dry and exhaling it saturated with water vapour at a partial pressure P_{EH_2O}. Then the portion of the total metabolic rate that is devoted to making good the loss of heat through evaporation in the respiratory tract is equal to $(P_{EH_2O} \cdot L)/(P_{ECO_2} \cdot J)$, where P_{ECO_2} is the mean carbon dioxide tension in the expired air and J is the energy released in cellular respiration per mole of carbon dioxide. Air is typically expired at a temperature of about 32–33 °C, so we may take P_{EH_2O} as 37.7 mmHg (the water vapour pressure at 33 °C). In addition, we may take L as 10.4 kcal/mol of water (43.7 kJ/mol) and J as 135 kcal/mol of carbon dioxide (562 kJ/mol), derived from Table 2.1. Thus the proportion of the total metabolic rate devoted to replacing the lost heat is $2.9/P_{ECO_2}$.

4.5.3 **On this basis, what proportion of the mammal's metabolism would be devoted to replacing the heat lost in the evaporation of water into dry inspired air if the mean P_{CO_2} of the expired air were to be (a) 2.9 mmHg and (b), more realistically, 29 mmHg?**

The conclusion is clear, even though we have ignored the additional heat needed to warm the air and the work required to ventilate the lungs. It should also be remembered that P_{CO_2} would be higher in the blood than in the expired air. All of these facts strengthen the conclusion that a mammal inhaling dry air and exhaling saturated air at 33 °C would be unable to maintain P_{CO_2} as low as in most fish.

As already noted, the expired air is usually below the core body temperature. Indeed, if air is heated in the nasal passages, and water evaporates there, then the surfaces must surely be cooled somewhat, in our case typically to 32–33 °C, but lower in a very cold environment. Many mammals make use of countercurrent heat exchange in their nasal passages to cool the expired air even more than in ourselves; less water vapour is lost in these animals – and less of the heat associated with evaporation. In the short term, however, heat loss is sometimes useful. A dog breathes mostly through its well-known cold nose and exhales air at about 29°C, but when it pants it exhales much warmer air through its mouth. Some birds reduce their arterial P_{CO_2} to one-third of the normal value during panting.

4.6 Water loss in expired air

The water vapour pressure in saturated air is 47 mmHg at 37 °C, 37.7 mmHg at 33 °C, and only 15 mmHg at 17.5 °C. Such values may be used to calculate the rates of water loss in the expired air under different conditions. The temperatures of 33 °C and 17.5 °C are intended here as representative, respectively, of expired air and of a moderately warm environment.

Consider the case of someone who is breathing in dry air and breathing out moist air at a rate of 15 000 l (at STP)/day (corresponding to a respiratory minute volume of 10.4 l/min). The atmospheric pressure is 760 mmHg (i.e. standard pressure) and corrections for temperature and pressure are not required.

4.6.1 **What is the daily rate of water loss (in g) if the air is inspired completely dry and expired saturated with water vapour at (a) 37 °C and (b) 33 °C?**

1 mol (18 g) of water vapour at STP occupies 22.4 l. Therefore, 1 l at STP weighs 0.80 g.

4.6.2 **From these two answers, how much water is saved daily by breathing out air at 33 °C rather than at 37 °C?**

4.6.3 **What is the daily rate of water loss if the inspired air is saturated with water vapour at 17.5 °C and the expired air is saturated at 33 °C?**

A typical rate of water loss in expired air is generally said to be about 400 ml/day. Small mammals have much higher relative metabolic rates than we do (Section 2.7) and they breathe correspondingly more rapidly. For desert rats it is therefore particularly worthwhile to cool the expired air.

4.7 Renewal of alveolar gas

This section makes use of normal lung volumes (tidal volume, dead space, etc.) to explore the renewal and stability of the alveolar gas. At the end, the opportunity is taken to introduce an example of an exponential time course and to point out some of its properties.

Fig. 4.2 is a conventional diagram illustrating, for a particular individual, lung volumes during normal breathing and with maximal inspiratory and

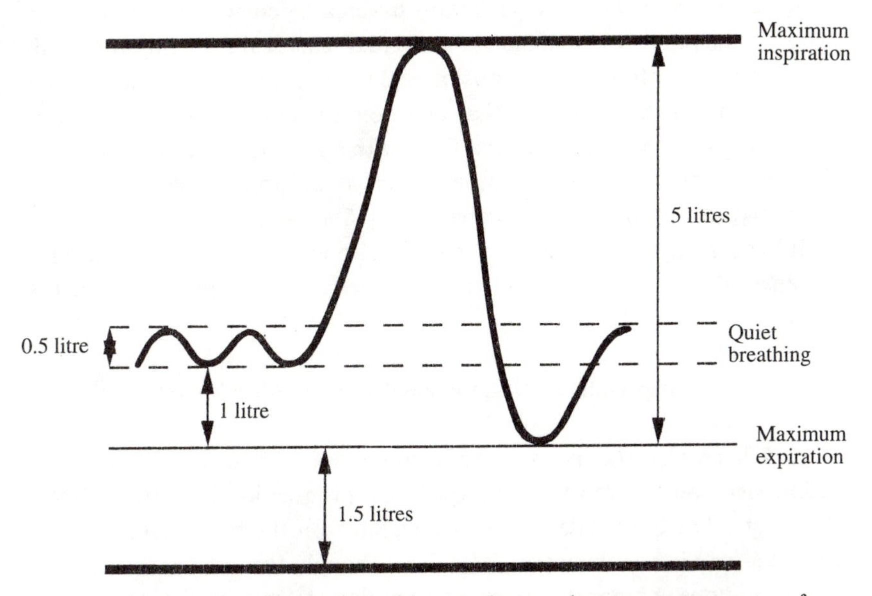

Fig. 4.2. Conventional illustration of lung volumes: the curve represents a few 'quiet' breaths followed by maximum inspiration and maximum expiration.

expiratory effort. ('Lung volumes' are usually regarded as including all airways.) The tidal volume during normal breathing is shown as 500 ml; about 150 ml of this represents dead space and 350 ml represents alveolar ventilation.

Think of this individual breathing regularly with the illustrated tidal volume of 500 ml. At the end of a normal expiration, 2500 ml of alveolar air remain in the alveoli and airways (the 'functional residual capacity'). In the subsequent inspiration this volume is mixed with 350 ml of inspired air to give a total of 2850 ml. The dilution of inspired air by alveolar air is therefore 350/2850 = 0.12 (i.e. about one-eighth).

4.7.1 **What proportion of the original alveolar gas remains in the alveoli after this one breath?**

That so much remains is important because of the resulting stability in alveolar gas tensions. If all the alveolar gas were renewed in every breath, there would be large fluctuations in alveolar P_{O_2} and P_{CO_2} and these would be communicated to the blood. It is therefore important to have a large functional residual capacity and residual volume. (It is also true that half-filled lungs need the least muscular effort in breathing.)

We now go on to consider breath-by-breath renewal of the alveolar gas, but the points to be made are more mathematical than physiological. As breathing continues, gases must in reality be exchanged between alveoli and blood, but let us ignore that fact now for mathematical simplicity, supposing that the lungs start out filled with a gas that is totally insoluble. For practical purposes helium is regarded as approximating to that property, for its solubility in water is only about 28% that of oxygen.

It has already been calculated that about one-eighth of the alveolar gas is renewed in each breath, so that after one breath the proportion left is 0.88.

4.7.2 **What proportion is left unrenewed after another breath?**

Fig. 4.3 graphs the progressive changes over further breaths, with successive points given by the geometrical progression 0.88, 0.88^2, 0.88^3, 0.88^4, etc. The time scale has been calculated on the basis of 12 breaths/minute (i.e. 1 breath every 5 s). Even though the change is actually discontinuous, a smooth curve has been drawn through the points. The curve is exponential, with an asymptote at zero. It has the following

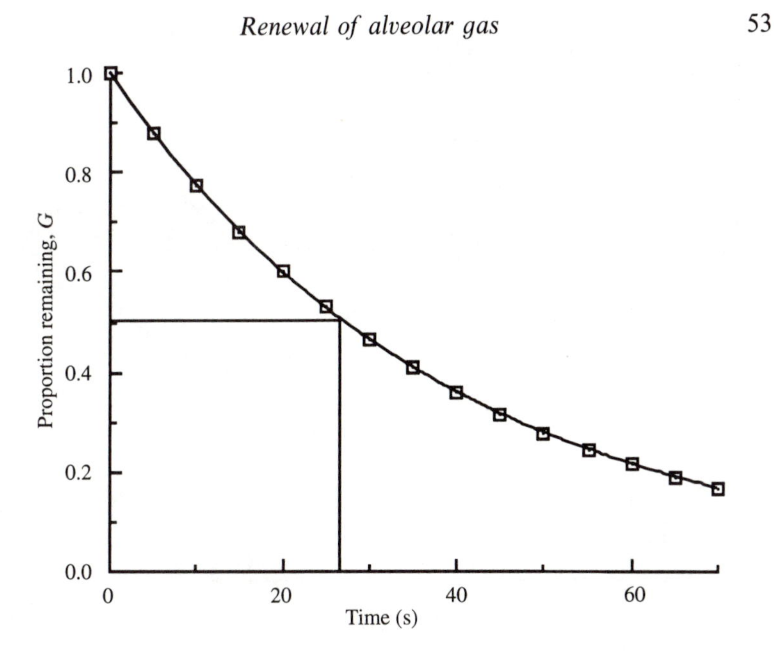

Fig. 4.3. The progressive renewal of alveolar gas over a period of 70 s at a breathing rate of 12/min. G is the proportion of the original volume remaining in the lungs. It is reduced by a factor of 0.88 at each breath. By means of the horizonal line at $G = 0.5$, and the vertical line through its intercept with the curve, the half-time, $t_{1/2}$, may be read from the time scale.

equation:

$$G = e^{-kt}, \tag{4.15}$$

where G is the proportion of gas remaining, t is the time in seconds and k is the 'rate constant' (equal here to $0.0256\ s^{-1}$, calculable as $-\ln 0.88$ divided by 5 s, the time per breath).

The time required for G to fall from any value to half that value is constant and is known as the half-time ($t_{1/2}$). Here it is 27.1 s. It may be calculated as $(\ln 2)/k = 0.692/k$.

If $\log G$ is plotted against t, a straight line results (Fig. 4.4). Its equation is:

$$\log G = -kt/(\ln 10) = -0.0111t. \tag{4.16}$$

The equation may also be written so as to feature the half-time instead of k:

$$\log G = -\log 2 \times t/t_{1/2}. \tag{4.17}$$

4.7.3 **As a check, what is the value of G, according to this equation, when $t = t_{1/2}$?**

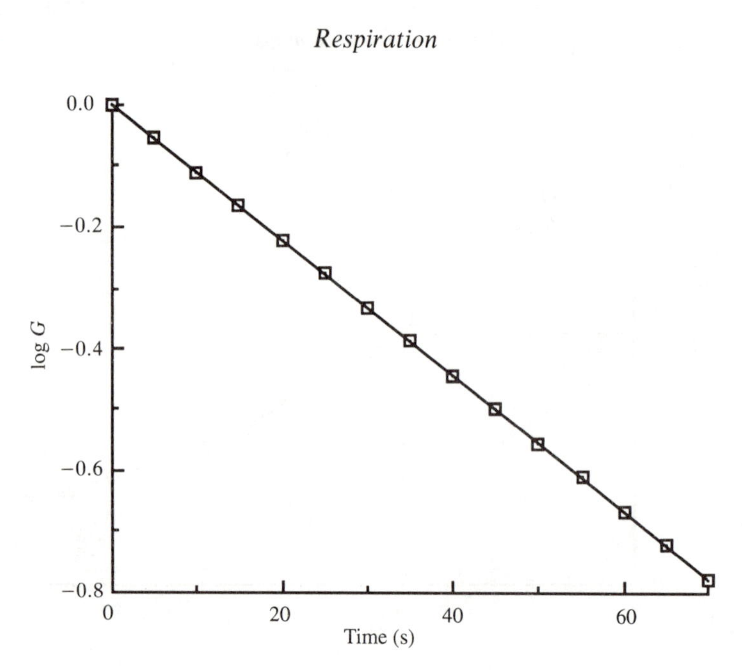

Fig. 4.4. Here the curve of Fig. 4.3 is straightened out by taking the logarithm of G, the proportion of the original alveolar gas remaining in the lungs.

Curves like those in Figs 4.3 and 4.4 may be obtained in tests of pulmonary function that involve the breathing of helium or pure oxygen. During the breathing of pure oxygen, the nitrogen initially present in the lungs is gradually lost and, if the expired air is analysed, 'nitrogen washout' or 'nitrogen clearance' curves may be obtained. These are normally linear when plotted as in Fig. 4.4. Departures from linearity occur when parts of the lungs are poorly ventilated.

4.8 Variations in lung dimensions during breathing

As lung volumes change during breathing (Fig. 4.2), so too do alveolar surface areas and linear dimensions. As a consequence, there are variations in the degree of stretch of elastic fibres, the curvature of the alveolar walls, and the state of the alveolar surfactant film.

4.8.1 **During quiet breathing as indicated in Fig. 4.2, what is the ratio of maximum to minimum lung volume?**

4.8.2 **In the same circumstances, if the lung spaces were to change in volume but not in shape, what would be the average variation in linear**

dimension, expressed as the ratio of maximum to minimum length?
($\sqrt[3]{1.2} = 1.063$)

Is this roughly as you imagined? The answer might be expected to apply (if only approximately) to the elastin fibres in the alveolar walls and indicates the variation in their degree of stretch. The next question concerns surface area – the area for gas exchange and the area covered by surfactant.

4.8.3 **In the same circumstances, what would be the average variation in surface area expressed as the ratio of maximum to minimum?**
($1.063^2 = 1.13$)

The point is not the exact answer, but the rough magnitude of the change. As alveoli become inflated, they do actually change in shape and appearance in ways that defy such simple mathematical treatment.

Consider next the thickness of the surfactant bearing film of fluid lining the alveoli. Since it is thin, its volume can be thought of as equal to the product of its surface area and its average thickness.

4.8.4 **Assuming that the surface area of this film increases during inspiration by 10%, and that its volume remains constant, by what percentage does its thickness decrease?**

4.9 Lung structure – branching of the airways

Here we consider the number of alveoli in the lungs and the amount of branching of the airways that is required to supply them.

The number of alveoli can be calculated as their total volume divided by the average volume of the individual alveoli. Both vary during breathing. For the representative individual of Fig. 4.1, the total volume of gas in the pair of lungs during quiet breathing is 2.5–3 l. From this should be deducted the amount in the airways. This is less than the total anatomical dead space, since part of this is outside the lungs, and the dead space is only about 0.15 l. Given the imprecision of the estimate of total lung volume, the correction is not worth making; most of the air is in the alveoli. Indeed, we can take any reasonable volume for the alveolar gas that is compatible with Fig. 4.1. Let us use 2.7 l. What is harder to be sure of is the corresponding average alveolar volume. If the alveoli are taken as roughly spherical, then their volumes can be estimated from their diameters, but estimating the average diameter is also problematical. Let us take it as 0.25 mm. This happens to be about the diameter found by the Revd. Stephen Hales in 1731, but it is actually based on more recent studies. The volume of a sphere is $4\pi/3 \times (\text{radius})^3$, or about half the cube of the diameter.

4.9.1 **Taking the alveoli as spheres of diameter 0.25 mm, what is the average volume of an alveolus, in mm^3 (µl)?**

4.9.2 **If the volume of alveolar gas is 2.7 l, how many such alveoli would there have to be?**

Is this the order of magnitude you expected? The large number of alveoli is the result of repeated branching of the airways (bronchi, bronchioles, alveolar ducts). The branching is mostly dichotomous, so that one airway branches into two narrower ones. The different stages in this branching are known as generations and these are commonly numbered in such a way that the trachea is '0', the two bronchi branching from it and leading to the lungs are '1', and so on (Fig. 4.5). The next question concerns the total number of generations needed to produce the estimated number of alveoli.

For the moment, let us consider a much simplified model of the lung; it is wrong in important details, but still instructive. In this model, the airways branch dichotomously, producing the same number of generations in all parts of the lung. The alveoli constitute the final generation. The number of airways increases progressively at each branching, according to the geometric progression 2, 4, 8, 16 and so on. The number at generation n is 2^n. Table 4.1 gives some values of n and 2^n.

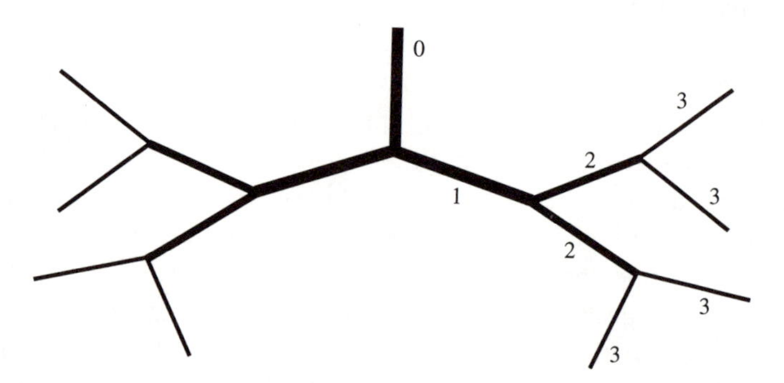

Fig. 4.5. An example of dichotomous branching, with generations of branches numbered as for the lungs. Generation 0 would be the trachea.

Table 4.1 *Selected values of* n *and* 2^n

1	2
2	4
12	4096
23	8 388 608
24	16 777 216
25	33 554 432
26	67 108 864
27	134 217 728
28	268 435 456
29	536 870 912

In relation to the dichotomous branching of airways, n is the generation number and 2^n is the number of airways of that generation.

4.9.3 **How many generations of branching, to the nearest round number, are required to explain the calculated number of alveoli in the two lungs?**

Evidently a huge number of alveoli can be achieved by comparatively few branchings. The number arrived at is not accurate, but note that it is not very sensitive to errors in the two quantities assumed in calculating the number of alveoli. Changing either one of these by a factor of two changes the estimated number of generations by only one, i.e. less than 4%.

The imperfections in this model of the lungs may be obvious from knowledge of histological structure, but it is instructive to explore the inadequacy of the model – both as a representation of the human lungs and as a design for the lungs of a hypothetical animal.

Assuming dichotomous branching, if the number of alveoli is A, there must be $A/2$ airways in the final generation of airways supplying them, $A/4$ in the generation before that, and so on. The total number of airways is thus the sum of the series:

$$A(1/2 + 1/4 + 1/8 + 1/16 \cdots).$$

The limiting value of the sum is equal to A; in other words, the total number of airways of all generations that supply the alveoli is equal to the number of alveoli themselves.

4.9.4 **For the model, what, approximately, is the ratio of the number of airways to the total number of airways plus alveoli?**

Now let us consider the implications of this for the dead space of the model. For this we need to specify an average airway volume. In reality the airways vary in volume, from trachea to bronchiole, but let us postulate simply that the average airway volume is very roughly like the volume of an alveolus. (Is this reasonable?) Then the total volume of gas in the model lungs would have to be largely dead space (i.e. about half). This is so far from realistic that it is scarcely worth quibbling over the choice of average airway volume. (How the alveoli would be packed within the lung is another question!)

In real lungs the dead space is a much smaller part of the total (see above) and this is largely because the alveoli are not arranged as in the model. The terminal bronchioles often give rise to three generations of respiratory bronchiole (which, by definition, have alveoli opening from their walls) and these lead via alveolar ducts to alveolar sacs, both of which are completely lined by alveoli. On average, there are said to be 13 alveoli opening from each alveolar duct or sac and this reduces by three or four the number of generations that are required to yield our 300 000 000 alveoli.

What does this imply about the anatomical dead space? In the model the number of airways supplying the alveoli equals the number of alveoli themselves. Suppose now that the model is made more realistic by having four fewer generations of airways, but the same number of alveoli.

4.9.5 **By what factor will the alveoli now outnumber the airways?**

The fewer airways means a much reduced, and more realistic, anatomical dead space. Without more information we cannot accurately estimate its size, but if we divide by 16 our earlier, unrealistic estimate for the model – of about half the total volume – then the resulting estimate of dead space is not unreasonable.

There is actually no exact answer for the number of generations. The reason is simply that there is not the same number in all parts of the lung. In some regions, alveoli appear after fewer generations than average and this means that there must be more generations elsewhere to compensate. Those without a special interest in the matter may be content to think of the generations as numbering 'about two dozen'.

4.10 Surface tensions in the lungs

The surface tension of the fluid lining the alveoli is important for its tendencies both to cause their collapse and to draw in fluid from the interstitial fluid and capillaries. According to Laplace's formula relating wall tension (T), internal pressure (P) and radius (r) for a sphere,

$$P = 2T/r. \tag{4.16}$$

The sphere might be, for instance, a bubble (with wall tension being surface tension) or an idealized spherical alveolus. Many books give this formula without specifying units; given as above without any numerical factor, the units are SI. Thus, if T is in N/m and r is in m, then P must be in N/m^2. Surface tensions are commonly given as mN/m (= dyne/cm), but a physiologist may be happier working with pressures in terms of mmHg or cm water. In dealing with alveoli, units of μm are more convenient than units of m. Here is the equation in such mixed units:

$$P \text{ (in mmHg)} = 15T \text{ (in mN/m or dyne/cm)}/r \text{ (in μm)}. \tag{4.17}$$

4.10.1 **Consider an air bubble in pure water with a surface tension of 70 mN/m. The radius of the bubble is 100 μm. What is the pressure, in mmHg, required to balance the surface tension?**

The radius was chosen as being a round number fairly representative of human alveoli, although these vary both from one to another and with degree of inflation. However, the fluid that lines the alveoli has a substantially lower surface tension than 70 mN/m. This is because of the presence of surfactant. The exact value of the alveolar surface tension varies during the respiratory cycle in a manner that is complicated by hysteresis. In other words, surface tension and surface area increase and decrease together, but the precise relationship between them varies with the direction of change. Generally the surface tension is between 5 and 30 mN/m. At equilibrium the surface tension is 25 mN/m. (Sometimes the surface tension in the alveoli is described as zero, but this is impossible.)

4.10.2 **Consider another air bubble, in saline and with the surface tension lowered by the action of lung surfactant to 25 mN/m. The radius of the bubble is again 100 μm. What is the pressure, in mmHg, balancing the surface tension?**

Although an analogy between a bubble and an alveolus is implied here, an alveolus is neither perfectly spherical nor even a completely enclosed space. Moreover, it is not kept inflated by the internal pressure (which fluctuates either side of zero during breathing), but is held open by the surrounding tissues (mainly other alveoli). The pressure just calculated for the surfactant-coated bubble must therefore signify something different in relation to an alveolus. It can be thought of as a negative pressure within the thin film of fluid lining the alveolus, and one that is tending both to collapse the alveolus and to draw more fluid through the alveolar epithelium from the interstitial fluid and capillaries. We return to this in the next section.

4.11 Pulmonary lymph formation and oedema

The filtration of fluid through capillary walls, whether systemic or pulmonary, depends on the balance of pressures indicated in Fig. 4.6, as first proposed by E. H. Starling towards the end of the last century. Both the hydrostatic pressures and the colloid osmotic pressures are usually higher inside the capillaries than out. The colloid osmotic pressure is typically lower in the interstitial fluid than in the plasma because plasma proteins are largely retained in the capillaries during filtration of fluid through the endothelium. The interstitial hydrostatic pressure is generally lower than the capillary blood pressure, and may be negative. The picture is clearest and most familiar in relation to the systemic capillaries and in these the net pressure is usually outwards near the arterial ends, but inwards towards the venules. On average within the body there is a net outwards pressure from the systemic capillaries and therefore a net outwards flow of fluid that returns to the circulation as lymph.

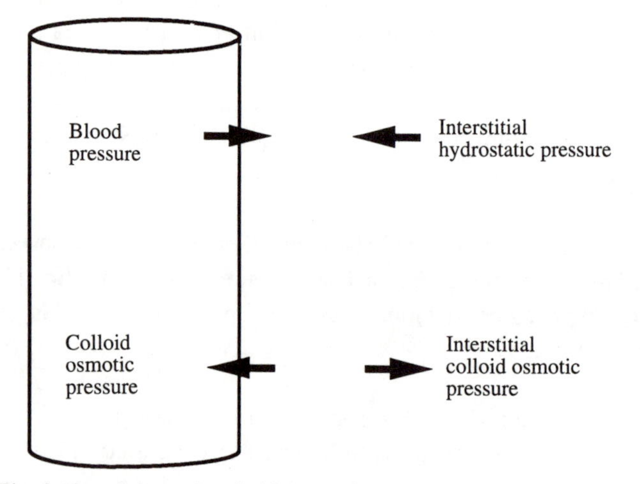

Fig. 4.6. The hydrostatic and colloid osmotic pressures influencing the flow of fluid across a capillary wall.

All the quantities used in this section are merely representative and some are hypothetical and difficult, or impossible, to measure. In most cases the natural variability must in any case be considerable. The calculations apply to parts of the lung at heart level; in a standing person the pressure in the alveolar capillaries might be about 15 mmHg higher at the base of the lungs through the effect of gravity, but pressure gradients along the capillaries are not considered here.

In old textbooks, from the 1950s for example, one might find the colloid osmotic pressure given, correctly, as around 25 mmHg and the average blood pressure in the pulmonary capillaries, at heart level, as about 7 mmHg, but with the two other pressures indicated in Fig. 4.6 being taken as negligible.

4.11.1 **On that basis, what is the net pressure difference across the capillary endothelium? In which direction would fluid flow?**

It was assumed in these older accounts that pulmonary oedema would occur when alveolar capillary blood pressure exceeded colloid osmotic pressure; only when pulmonary blood pressures are abnormally high would there be any obvious role for the pulmonary lymphatic vessels. It is now apparent that the situation is more complicated and more interesting. The space between the alveolar capillaries and epithelium is so small that their basement membranes are contiguous. The hydrostatic pressure there is negative, with a value estimated as about −9 mmHg. The three pressures so far given are shown in Fig. 4.7. As for the interstitial colloid osmotic pressure, it cannot be measured directly, and even the more bulky pulmonary lymph that drains from it is not easily accessible.

4.11.2 **What would the colloid osmotic pressure of the interstitial fluid have to be for the various pressures just to balance out?**

This is 36% of the plasma value of 25 mmHg and is in general accordance with the assumption of partial filtration of protein. For lymph to be produced, only a slightly higher value would be required and this would still be credible. In fact the concentration of protein in the pulmonary lymphatics is roughly half that of blood plasma. A round value of 10 mmHg would seem to be consistent with lymph flow and this value is used below.

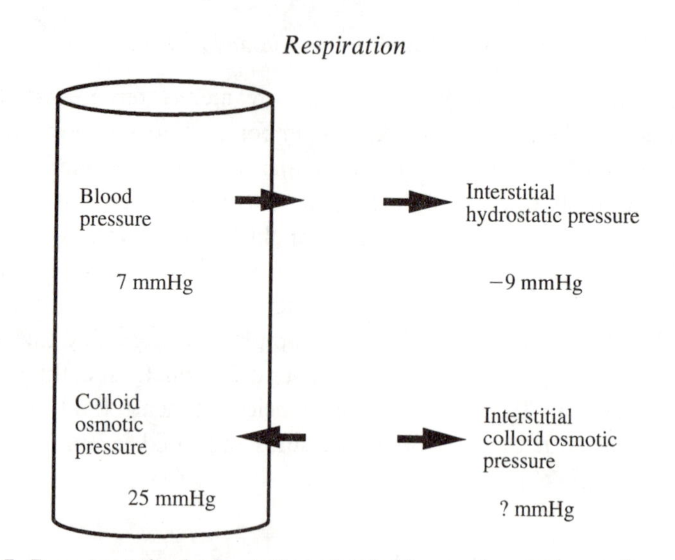

Fig. 4.7. Representative hydrostatic and colloid osmotic pressures influencing the flow of fluid across the wall of an alveolar capillary.

'Pulmonary oedema' is usually taken to mean the accumulation of excess fluid, not in the interstitial fluid, but in the alveolar air-spaces. There is normally a very thin layer of fluid lining these spaces and this bears in turn an even thinner surface film of surfactant. Because this has a surface tension, albeit substantially lower than that of a simple salt solution, there is in the film the equivalent of a negative hydrostatic pressure (Section 4.10). The actual pressure depends on the surface tension and local radius of curvature in accordance with the Law of Laplace. A value of nearly −4 mmHg was calculated in 4.10.2 for an alveolus-like bubble of radius 100 μm, but here let us try another approach.

The air pressure in the alveoli fluctuates by about 1 mmHg either side of zero during quiet breathing and may be neglected here. In the interstitial fluid there is, using values from above, a negative hydrostatic pressure of 9 mmHg and a colloid osmotic pressure of 10 mmHg, both of which tend to draw water from the alveoli. Opposing these pressures is the colloid osmotic pressure of the alveolar fluid. This has not been measured, but it has been argued that it would tend to approximate that of the interstitial fluid. (The argument, in brief, is that protein moves across the alveolar epithelium by diffusion and pinocytosis and reaches about the same concentration either side.) If the volume of surface film is assumed to be steady, the hydrostatic pressure in the surface film is calculable from the balance of hydrostatic and colloid osmotic pressures. It is marked 'x' in Fig. 4.8.

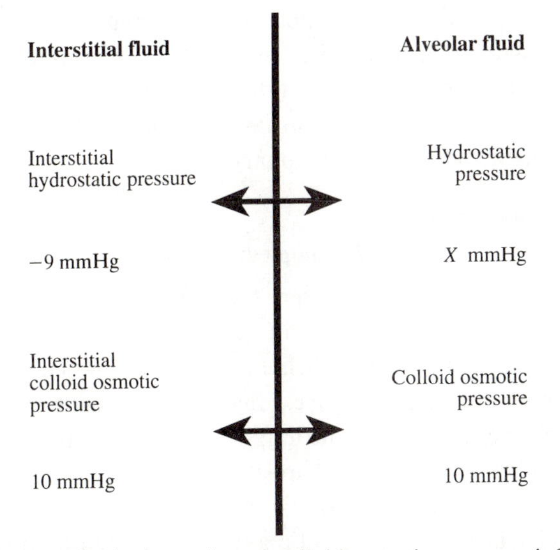

Fig. 4.8. Representative hydrostatic and colloid osmotic pressures influencing the flow of fluid between an alveolar capillary and the fluid lining an alveolus.

4.11.3 **If the hydrostatic pressure in the interstitial fluid is −9 mmHg and the colloid osmotic pressures of the interstitial fluid and surface film are both 10 mmHg, what is the hydrostatic pressure in the surface film?**

This negative pressure must be due to surface tension. The value of the latter can be calculated if a value is assumed for the radius of curvature of the surface film. An obvious starting point is the radius of a typical alveolus – 120 μm perhaps. Here is equation (4.17) again, giving Laplace's formula for a sphere in convenient units:

$$P \text{ (in mmHg)} = 15T \text{ (in mN/m or dynes/cm)}/r \text{ (in μm)}. \quad (4.17)$$

4.11.4 **For the film lining an (idealized) alveolus of radius 120 μm, what is the surface tension that is equivalent to a negative hydrostatic pressure in the surface film of 9 mmHg?**

This is close to the surface tension of pure water at body temperature (i.e. about 70 mN/m) and far too high for alveolar fluid with surfactant (for which the equilibrium value is 25 mN/m). This suggests that the calculation is based on a wrong assumption. When an alveolus is viewed in section, the air–fluid interface is seen to be very far from smoothly spherical. Rather, its curvature varies locally, being sometimes concave and sometimes convex. The radius of curvature must be quite small in some of the nooks and crannies where the fluid tends to collect. If the surface film is to be stable and fluid is not to be drawn into the alveoli, the smallness of these radii must be offset by a lower surface tension. Surfactant is therefore essential to the avoidance of pulmonary oedema.

4.12 The pleural space

The whole pleural space, between the visceral and parietal pleurae, contains no more than about 2 ml of fluid. The elastic recoil of the lungs tends to separate the two mesothelial membranes and enlarge the space, making the pressure inside negative. Fluid therefore tends to enter through the permeable mesothelia and the lungs can be held against the chest wall only if there is an opposing mechanism for fluid removal. That must not be too effective, however, or lubrication would suffer.

We consider here the hydrostatic and colloid osmotic pressure differences (the Starling forces) between pleural space and capillary blood, doing so separately for the parietal and visceral sides. Representative pressures are summarized in Fig. 4.9. Note, however, that they represent a simplification of the true situation, for the negative inside pressure not only shows a vertical gradient as in any body of fluid, but fluctuates

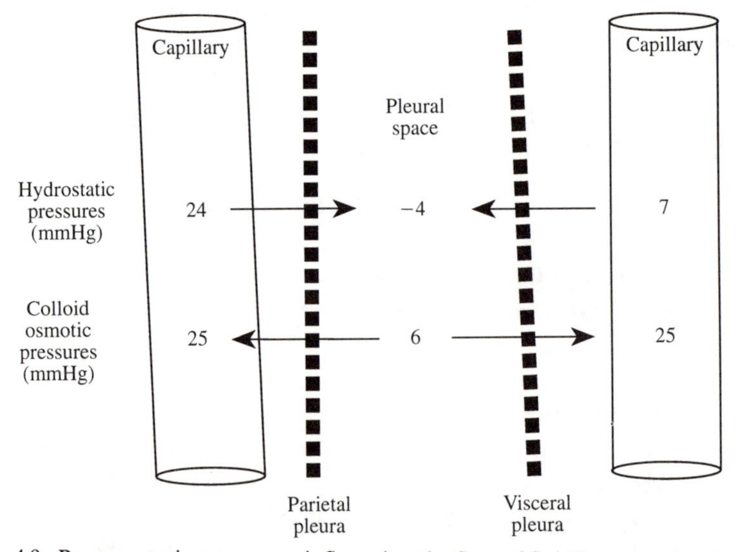

Fig. 4.9. Representative pressures influencing the flow of fluid between the pleural space and the capillaries separated from it by the parietal and visceral pleurae.

with breathing; here we simply take a mean pressure of −4 mmHg. As to the colloid osmotic pressure that tends to draw water into the space, this is shown as 6 mmHg. This is much lower than in plasma and corresponds to a protein concentration that is only a quarter or fifth of the plasma value.

Parietal pleura

In the (systemic) capillaries close to the parietal pleurae, the mean blood pressure at heart level is some 22–26 mmHg. Let us take it as 24 mmHg. The colloid osmotic pressure of the blood plasma is typically near 25 mmHg. It is assumed for present purposes that there is no significant active transport generating any further osmotic gradient across the pleural mesothelium. Values for pressures in the interstitial fluid surrounding the capillaries are not required for the next calculation.

4.12.1 **What is the net pressure difference between systemic capillaries and pleural space, taking into account both hydrostatic and colloid osmotic pressures? In what direction does it act?**

Visceral pleura

A similar calculation may be carried out for the other side of the pleural space. The visceral pleurae are supplied with both bronchial (systemic) and pulmonary blood, but mainly with the latter. The blood pressure in the pulmonary arteries is much lower than in the aorta and the pressure in the pulmonary capillaries is therefore lower than in the systemic capillaries – let us say 7 mmHg at heart level (Fig. 4.9). The colloid osmotic pressure of the plasma may be taken again as 25 mmHg.

4.12.2 **What is the net pressure difference between pulmonary capillaries and pleural space, taking into account both hydrostatic and colloid osmotic pressures? In what direction does it act?**

One may question some of the exact pressures shown in Fig. 4.9, but the conclusions seem clear, that fluid tends to enter the pleural space on one side and leave on the other. There is thus one mechanism for keeping the pleural space small and another that ensures the presence of enough fluid for lubrication. The colloid osmotic pressure in the space must be kept low, or fluid removal would be prevented. Protein leaves the space by lymphatic drainage and pinocytosis, and some is carried out in the flow of fluid through the visceral pleurae (the process known as 'solvent drag').

5
Renal function

I start at the glomerulus – with the composition of the filtrate, the filtration rate, and the influence on the latter of colloid osmotic pressure; I hope the treatment in these sections (5.1 and 5.2) will be less familiar than the concepts. Moving from glomerular to tubular processes, I apply the clearance concept to inulin, urea and other substances and then go on to consider the effects of water reabsorption on the tubular concentrations of inulin and urea. In Section 5.6, quantification of the rates of filtration and reabsorption for sodium bicarbonate is used to bring out some useful generalizations concerning kidney function.

Calculations relating to apparently 'isosmotic' reabsorption and the mechanisms and energy requirements of sodium reabsorption (Sections 5.7 and 5.8) place the emphasis on the bulk movements of water and solutes, rather than on the fine tuning of urine composition. Homeostasis is not neglected, however. Thus, the renal regulation of extracellular fluid volume is approached via autoregulation of glomerular filtration rate and glomerulotubular balance (Sections 5.10 and 5.11) and there are calculations on the interrelationships amongst solute excretion, water excretion and urine concentration (Sections 5.12 to 5.14). Finally (Section 5.15), certain aspects of the medullary countercurrent system are tentatively quantified in an attempt to resolve points that are often left vague in elementary accounts.

5.1 The composition of the glomerular filtrate

Osmotic pressure

In 1844, Carl Ludwig suggested that the first stage in the formation of urine is the production of a protein-free ultrafiltrate of blood plasma, this being driven through the walls of the glomerular capillaries by the pressure of the blood. Fluid from Bowman's capsule has now been shown, by chemical analysis, to resemble such an ultrafiltrate; plasma and capsular fluid have virtually the same osmotic pressure. (For the small difference, see Section 6.7.) That the osmotic pressures should be similar was concluded as long ago as 1896, by G. Tammann, on the basis of what is essentially the following argument.

If a solution, such as blood plasma, is forced under pressure through a filter that holds back a proportion of the solutes, then the filtrate has a lower osmotic pressure. The difference in osmotic pressure between the filtrate and the original solution would not exceed the difference in hydrostatic pressure. If it did, then filtration would cease. To complete the argument one needs only to show that the osmotic pressure of the plasma is vastly greater than the difference in hydrostatic pressures.

The osmotic pressure of the plasma is most usually expressed in terms of the total concentration of solutes, i.e. its osmolarity or osmolality (Section 6.9). As a round number, the osmolality is 300 mosmol/kg water. The osmotic pressure in mmHg can be calculated from the osmolality by means of the following relationship (see Notes and Answers for Chapter 6):

$$1 \text{ mosmol/kg water} = 19.3 \text{ mmHg at } 37\,^{\circ}\text{C}. \tag{5.1}$$

5.1.1 **What is 300 mosmol/kg water in terms of mmHg at 37 °C?**

This answer needs to be compared with the hydrostatic pressure difference between plasma and filtrate in the renal corpuscles. Putting ourselves in Tammann's position of long ago, we would not know a typical value for this difference. It happens, however, that even a rough estimate will do. Assuming a mean arterial blood pressure of 100 mmHg, allowing for some fall in pressure between arteries and glomerular capillaries, and ignoring any back pressure in Bowman's capsule, let us take (with an eye to ease of calculation) a hydrostatic pressure difference of 60 mmHg between plasma and filtrate. This, we are assuming, represents the maximum conceivable difference in osmotic pressure between plasma and glomerular filtrate.

5.1.2 **What is this maximum difference in osmotic pressure between plasma and glomerular filtrate, expressed as a percentage of the plasma osmotic pressure as calculated in** 5.1.1**?**

The actual difference in osmotic pressure is equal to the colloid osmotic pressure of the plasma (Section 6.7). Larger differences in the concentrations of *individual* solutes are not ruled out by this argument, but actual differences are small, except where a substance (such as calcium) is bound to plasma proteins and is therefore held back with them. The Donnan effect (Section 6.6) produces only small ionic gradients.

Protein concentration

Textbooks may say, following Ludwig, that the glomerular filtrate is like the plasma from which it forms, except that it is free of protein. That the protein content is at least low is immediately evident in electron micrographs of glomeruli, in which the filtrate appears clear white and the plasma (if not removed by perfusion) appears grey. Students that have been taught, without qualification, that the glomerular filtrate is protein-free can be puzzled to learn that there is a mechanism in the brush borders of the proximal tubules for the recovery of filtered protein. So, we may ask, is the filtrate truly protein-free? Or, if it nearly is so, why should there be a mechanism for protein reabsorption? A quick calculation answers the second question.

For this, let us postulate that glomerular filtrate has a low content of protein, low enough that one might feel justified in calling the filtrate 'essentially protein-free'. The protein content of plasma is about 70 g/l, so an arbitrary postulated concentration of 0.01–0.1 g/l for the filtrate might seem to justify that phrase. (Measured concentrations do actually lie in that range, but it remains uncertain what values are typical of the healthy human kidney.)

5.1.3 **A typical glomerular filtration rate is about 180 l/day (125 ml/min). How much protein would the daily 180 l of filtrate contain if the concentration in the filtrate were 0.01–0.1 g/l?**

This filtered protein would be lost to the body if it were not reabsorbed. Moreover, it is not negligible as compared with the daily protein intake. The 'recommended daily allowance' of protein for a moderately active young man has been given as about 70 g. Clearly the recovery of the calculated daily amount of protein could be life-saving during starvation.

5.2 The influence of colloid osmotic pressure on glomerular filtration rate

Glomerular filtration is driven by the hydrostatic pressure of the blood in the glomerular capillaries and it is opposed both by the lesser

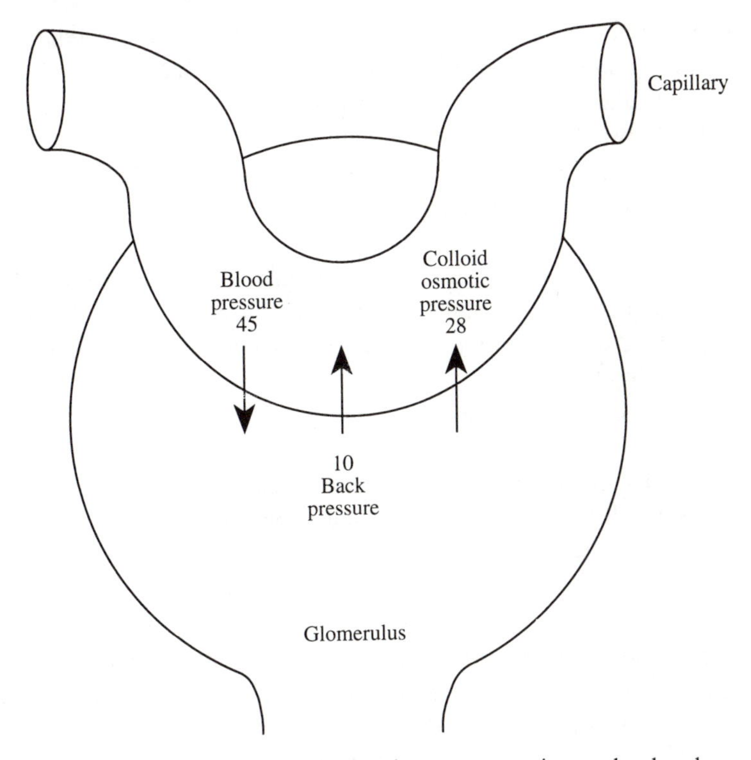

Fig. 5.1. A diagrammatic renal corpuscle showing representative textbook values for the mean hydrostatic and colloid osmotic pressures (in mmHg) that influence the rate of glomerular filtration. The capillary blood pressure is based on measurements made on rat kidneys.

back-pressure within Bowman's space and the colloid osmotic pressure of the plasma proteins in the blood. Fig. 5.1 indicates representative textbook values for each of these (but remember that the hydrostatic pressures vary from moment to moment, species to species and glomerulus to glomerulus).

The colloid osmotic pressure of the plasma rises along the length of each capillary as filtration of fluid concentrates the remaining protein; the value shown is a half-way value between afferent and efferent arteriolar blood. The rise in colloid osmotic pressure can be calculated from the 'filtration fraction', itself determined from glomerular filtration rate (GFR) and renal plasma flow (RPF):

$$\text{filtration fraction} = \text{GFR}/\text{RPF}. \tag{5.2}$$

The RPF could be, say, 660 ml/min (compatible with a renal blood flow of 1.2 l/min and a haematocrit of 45%) and the GFR 125 ml/min. These figures yield a filtration fraction of 0.19. This means that the protein entering the glomeruli in 1 ml of plasma leaves it in only 0.81 ml. The colloid osmotic pressure rises slightly more than the protein concentration, but for present purposes we may take the two as exactly proportional to each other.

5.2.1 **If the plasma entering the glomerular capillaries has a colloid osmotic pressure of 25 mmHg and the filtration fraction is 0.2, what is the colloid osmotic pressure in the plasma entering the efferent arterioles?**

The colloid osmotic pressure shown in Fig. 5.1 is midway between 25 mmHg and this value. The net filtration pressure is 7 mmHg (i.e. 45 – 10 – 28 mmHg). For a given permeability of the filtration barrier, the GFR is proportional to the mean net filtration pressure. The next calculation reveals how sensitive the GFR is to a change in colloid osmotic pressure.

5.2.2 **Given initial pressures as in Fig. 5.1, what is the percentage increase in net filtration pressure (and hence in GFR) if the mean colloid osmotic pressure falls (by 7%) to 26 mmHg?**

The point here is to demonstrate the magnifying sensitivity of GFR to colloid osmotic pressure rather than exact quantification of this change. A fuller treatment would take into account the rise in colloid osmotic pressure and fall in blood pressure along the length of each glomerular capillary, as well as possible changes in back-pressure within Bowman's capsule. There is also the possibility to be considered that there are compensating changes in glomerular blood pressure – the autoregulation of GFR as discussed in Section 5.10. Dilution of the plasma proteins by an infusion into the blood of isosmotic saline does lead to a prompt diuresis in accordance with the above effect, but much of this diuresis is due to a reduction in fluid reabsorption.

5.3 Glomerular filtration rate, inulin and PAH clearance, drug clearance

A typical glomerular filtration rate (GFR) for a 70 kg man is 180 l/day or 125 ml/minute. GFR is measured by the 'clearance' technique, using substances that are freely filtered at the glomeruli and then neither reabsorbed nor secreted by the tubules. An ideal substance for the purpose is the plant carbohydrate inulin. This has to be infused into the blood, since it is not naturally present. The technique makes use of the fact that, for inulin,

$$\begin{aligned} \text{rate of excretion} &= \text{rate of filtration} \\ &= \text{GFR} \times \text{concentration in plasma}. \end{aligned} \quad (5.3)$$

GFR is thus estimated as:

(rate of inulin excretion)/(concentration of inulin in plasma),

this ratio being the 'inulin clearance' – see Notes and Answers.

5.3.1 (Practice example) **A man is infused with inulin at a rate adjusted to maintain the plasma concentration at 4 mg/l. The rate of renal excretion is found to be 0.25 mg/min (360 mg/day). What is the GFR in ml/min or l/day?**

The answer is half the average normal value given above. Is the subject a small man? Has he lost one kidney? Has he a reduced number of functioning nephrons as a result of old age? GFR may be reduced to about 50% at the age of 80 and the reduction in the elderly is important in relation to drug dosage. If a drug is removed from the body by the kidneys and GFR is low, then its concentration remains high for a longer time.

Let us consider now the special case of a drug that is rapidly dispersed in the extracellular fluid and which is lost from that only by glomerular filtration. This particular drug does not enter the cells or other body compartments. It is neither reabsorbed nor secreted by the kidney tubules. It does not bind to plasma proteins. Its properties are thus much like those of inulin. The excretion rate is equal to the GFR multiplied by the concentration in the plasma (assumed equal to the concentration in the glomerular filtrate). As excretion proceeds, the plasma concentration falls and so too, therefore, does the rate at which the drug is filtered. The plasma concentration declines exponentially with time (Fig. 5.2). It is a property of such an exponential decline that the period required for the concentration to halve, i.e. the 'half time', $t_{1/2}$, is given by the following formula:

$$t_{1/2} = (\ln 2)/k = 0.693/k, \quad (5.4)$$

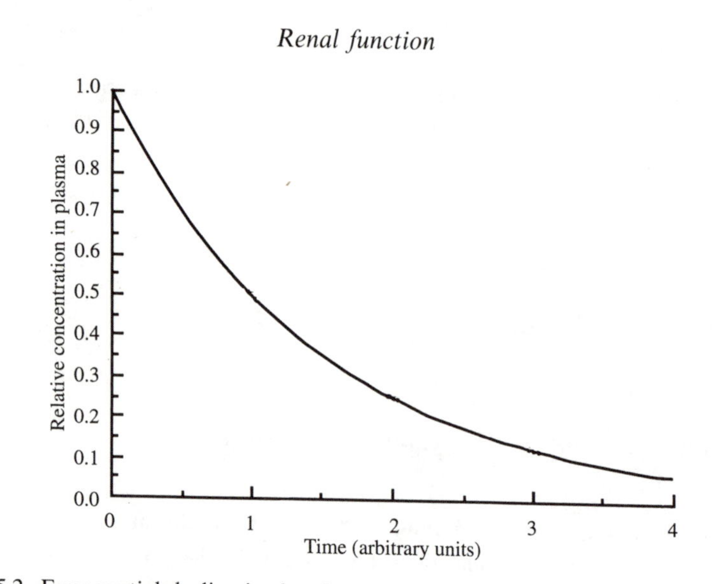

Fig. 5.2. Exponential decline in the plasma concentration of a hypothetical drug as it is lost from the extracellular fluid by glomerular filtration.

where k is the 'time constant' (see also Section 4.7). In this case,

$$k = \frac{\text{GFR}}{\text{extracellular fluid volume}}. \tag{5.5}$$

5.3.2 **How long does it take for the concentration of the drug (or inulin) to be reduced by half if the volume of extracellular fluid 14 000 ml and the GFR is (a) 125 ml/min, and (b) half that (62.5 ml/min)?**

From the answer to (a) it is obvious why, for the measurement of inulin clearance, it is necessary to maintain the plasma concentration by continuous infusion of inulin rather than simply giving a single injection.

5.3.3 **How long does it take for the concentration of the drug to be reduced by three-quarters (i.e. half plus a quarter) by excretion if the GFR is 125 ml/min?**

Rates of excretion must be less if the filtered drug is partially reabsorbed through the walls of the tubules (as is urea). Excretion must be faster if the drug is actively secreted into the tubules (as is uric acid). A substance well known to be transported from blood to tubular fluid (as well as being filtered) is para-amino hippuric acid (PAH). If we assume that all the PAH entering the renal arteries is excreted (as is nearly true), then we have the following relationship:

renal plasma flow × concentration of PAH in plasma

= rate of excretion of PAH

= rate of unit flow × concentration of PAH in urine. (5.6)

Thus the renal blood flow may be estimated as:

$$\frac{\text{rate of PAH excretion}}{\text{concentration of PAH in plasma}}.$$

This is the 'PAH clearance'. (In fact some of the blood, about 10%, by-passes the nephrons, so what is actually estimated this way is the 'effective renal plasma flow'.)

A typical textbook value for the renal blood flow is 1200 ml/min. Since about 55% of the blood consists of plasma, the renal plasma flow is about 660 ml/min.

5.3.4 **Assume now that a drug is completely cleared from the blood as it flows through the kidneys (as PAH nearly is). The renal plasma flow is 660 ml/min. The drug is confined to the extracellular fluid (of volume 14 000 ml) and is not bound to plasma proteins. Again the concentration in the plasma declines exponentially. What is the half-time?** k in equation (5.4) is now given by the renal plasma flow divided by the volume of extracellular fluid.

5.4 The concentrating of tubular fluid by reabsorption of water

By the end of the proximal convoluted tubule roughly two-thirds of the filtered fluid has been reabsorbed. The actual proportion depends on the individual nephron, the rate of filtration etc., and values in textbooks vary accordingly.

5.4.1 **Suppose that inulin has been infused into a person's blood – as for the measurement of GFR (Section 5.3). In the glomerular filtrate the concentration is as in the plasma water. Inulin is neither reabsorbed from the tubular fluid nor secreted into it. By what factor is the inulin concentrated by the end of the proximal convoluted tubule if two-thirds of the fluid is reabsorbed?**

Any filtered substance tends to be concentrated this way, though the tubular concentration may nevertheless fall because of reabsorption (as in the extreme case of glucose).

Fluid reabsorption continues along the nephron (though negligibly in the ascending thick limb of the loop of Henle) and only a tiny proportion of the filtered water emerges in urine.

5.4.2 **Suppose that the glomerular filtration rate is 125 ml/min and that urine is produced at a rate of 1.25 ml/min. By what factor is the inulin concentrated in the urine?**

It is no small point that inulin, a substance foreign to the body, should become concentrated in the urine in this way. Why does the vertebrate kidney filter and reabsorb so much fluid, if not to ensure excretion of substances, like inulin, for which no special transport systems have been evolved?

5.4.3 **The concentration of glucose in plasma is about 5 mmol/l. The renal reabsorption of glucose can be inhibited by phlorizin. With complete inhibition, what would be the concentration of glucose in the urine, assuming the flow rates as given in question 5.4.2?**

This value is quite high compared with the osmotic concentrations of both plasma (300 mosmol/kg water) and urine (30–1400 mosmol/kg water). An osmotic diuresis would result (as it does in diabetes mellitus, where glucose reabsorption is typically incomplete) and the concentrating of the glucose would therefore actually be less than has been calculated. (For further discussion of osmotic diuresis, see Section 5.13.)

5.5 Urea – clearance and reabsorption

It has not always been clear that the first process in urine formation is filtration at the glomeruli. Rudolph Heidenhain did not believe in filtration and, in 1883, revived instead an older theory of glomerular secretion. One reason for his views was the very high value for the glomerular filtration rate that was required to account for the known rate of urea excretion. This he calculated on the assumptions: (1) that urea is filtered at the same concentration as in plasma; and (2) that the rate of excretion is the same as the rate of filtration. In accordance with these assumptions, equation (5.3) should apply just as for inulin:

$$\text{rate of excretion} = \text{rate of filtration}$$
$$= \text{GFR} \times (\text{concentration in plasma}). \qquad (5.3)$$

The normal concentration of urea in plasma averages about 4.5 mmol/l (270 mg/l) with a range of about 2.5–7.5 mmol/l. The excretion rate is typically about 270–580 mmol/day.

5.5.1 **On the above assumptions, what GFR (in l/day) would correspond to an excretion rate of 450 mmol/day and a plasma concentration of 4.5 mmol/l?**

Using his own figures, Heidenhain obtained a roughly similar estimate for GFR (70 l/day). Although we now know that the true GFR (for a man of 70 kg) is typically close to 180 l/day, Heidenhain regarded his estimate as too preposterously high to be real. That his estimate was actually far too low results from the fact that much of the filtered urea is reabsorbed into the blood – a fact that surprises those who do not expect reabsorption of an excretory product.

What was effectively calculated in question 5.5.1 is what is known as the 'urea clearance', though this is usually expressed in ml/min. The actual urea clearance depends on the rate of urine flow, being least when the flow of urine is slow and rising with increasing flow rates (Fig. 5.3). The answer to 5.5.1 (equivalent to almost 70 ml/min) is typical of flow rates exceeding 2 ml/min.

We have seen that the urea clearance provides an underestimate of glomerular filtration rate, this being because much of the filtered urea is reabsorbed. The proportion that is reabsorbed can be calculated from the following formula; it applies to any substance that is filtered at the same concentration as it occurs in plasma.

$$\text{fraction reabsorbed} = 1 - (\text{clearance}/\text{GFR}). \qquad (5.4)$$

5.5.2 **If the urea clearance is 70 ml/min and the GFR is 125 ml/min, what fraction of urea is being reabsorbed?**

As to the mechanism of reabsorption, urea, like inulin (Section 5.4), is concentrated in the tubules by the reabsorption of water. The resulting concentration gradient between lumen and surrounding blood leads to

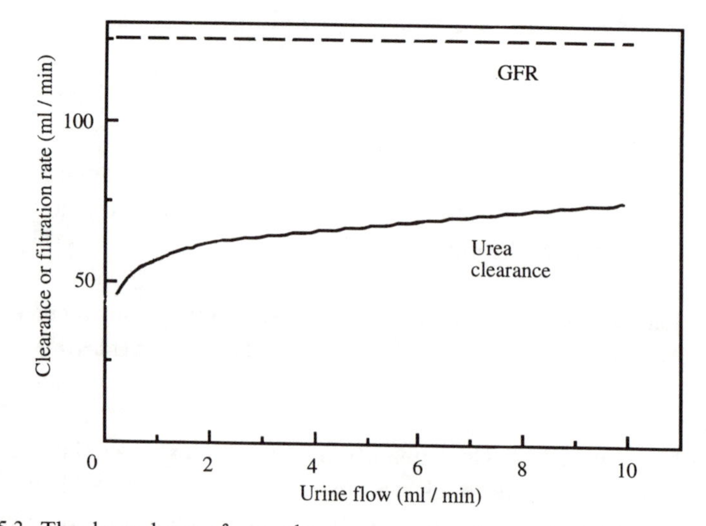

Fig. 5.3. The dependence of urea clearance on rate of urine flow. The broken line shows a constant glomerular filtration rate of 125 ml/min.

outward diffusion through the permeable walls of the tubules. As a result, the concentration of urea in the urine is elevated less than it would otherwise be. Let us consider now only what happens in the proximal convoluted tubules, for beyond these the situation is complicated by the medullary concentration gradients that are so important to the counter-current mechanism. In the proximal convoluted tubules the maximum conceivable proportion of filtered urea that could be reabsorbed by the passive mechanism just indicated (i.e. the limiting value) would correspond to virtually zero elevation of urea concentration compared with plasma.

5.5.3 **What is this maximum proportion of urea (i.e. the limiting value) that could be reabsorbed in the proximal tubules by this mechanism if two-thirds of the filtered water is reabsorbed there?**

Compare this with the answer to 5.5.2.

In the event of total kidney failure, urea excretion ceases, but the hepatic production of urea continues. Urea therefore accumulates in blood and tissues.

5.5.4 **Suppose that urea is produced within the body at a rate of 450 mmol/day following complete renal failure (so that none is excreted) and that it is distributed evenly through all compartments of body water. If the volume of this water is 45 l, by how much does the concentration of urea rise in one day (in mmol/l)?**

5.6 Sodium and bicarbonate – rates of filtration and reabsorption

It is important to realize just how much sodium or sodium chloride is reabsorbed, and the first two calculations are about this. Consideration of bicarbonate reabsorption and acid excretion then leads to important generalizations about kidney function.

The rate at which any solute is filtered may be calculated as the product of the GFR and the concentration of the solute in the filtrate. The latter, for anything but protein and substances strongly bound to protein, may be taken as approximating the concentration in blood plasma. (Small differences due to the Donnan effect are discussed in Section 6.6.) For the concentration of sodium in the filtrate a rough value of 150 mmol/l will do; if this seems a little high for typical plasma, remember that the sodium there is diluted somewhat by the volume of plasma proteins (Section 6.8). As already noted, the glomerular filtration rate (GFR) of a 'textbook' man of 70 kg averages 180 l/day.

5.6.1 **According to these values, how many moles of sodium are filtered in a day?**

Expressed thus, in terms of moles, the quantity is hard to visualize, so it may help to think in terms of a mound of salt. The filtered sodium is accompanied mainly by chloride and, to a lesser extent, bicarbonate, but for the purpose of visualization we may consider it all as sodium chloride. The formula weight of NaCl is 58.5.

5.6.2 **Expressed in terms of kg of its chloride, how much sodium is reabsorbed by the kidneys in a day?**

The 27 mol of sodium filtered in a day compare with a daily excretion, by people on a typical western diet, of 0.08–0.2 mol. These quantities are so very different that 27 mol/day does equally well as an estimate of the rate of renal sodium reabsorption. Likewise, since only about 1 l of urine is produced per day, the value of 180 l/day for a typical GFR does just as well for the typical rate of fluid reabsorption.

What these values illustrate is that the dominant activity of the kidneys is not the selective excretion that tends to receive most attention, but the filtration and reabsorption of huge amounts of solute and water. We may explore this further with regard to bicarbonate reabsorption and acid excretion.

5.6.3 **Given a GFR of 180 l/day and a concentration of bicarbonate in the filtrate of 25 mmol/l, what is the rate of bicarbonate filtration in mmol/day?**

Human urine is usually acid and contains little of this filtered bicarbonate. One can thus say of bicarbonate, as of sodium, that the rate of reabsorption approximates to the rate of filtration. The various transport mechanisms in the kidneys involve many kinds of solute, many kinds of channels and carriers, and many cell types. It is perhaps helpful at times, therefore, to realize that the main task of the tubules is to reabsorb a fluid that is, in essence, a glomerular filtrate, or (almost) deproteinated blood plasma. Furthermore, these fluids approximate to a solution of NaCl and $NaHCO_3$. This means that chloride and bicarbonate must be reabsorbed in about the same ratio as they occur in plasma.

Some of the filtered bicarbonate passes through the walls of the tubules as bicarbonate ion. Mostly, however, it is decomposed within the tubules by secreted hydrogen ions and then reconverted to bicarbonate in the epithelial cells. Usually more hydrogen ions are secreted than are needed to effect bicarbonate reabsorption and there is then acid *ex*cretion. Most of the excreted hydrogen ions are combined with ammonia (forming ammonium ions) and with phosphate and other buffers. It is important to appreciate how different the rates of acid excretion are from those of bicarbonate filtration and acid secretion. Acid excretion is usually equivalent to 40–110 mmol hydrogen ions per day (though it can be negative and the urine therefore alkaline).

5.6.4 **Suppose that acid is excreted at a rate equivalent to 90 mmol/day of hydrogen ions (i.e. at a high rate) and that the rate of bicarbonate filtration is as calculated in** 5.6.3. **What is the first rate as a percentage of the second?**

5.7 Is fluid reabsorption in the proximal convoluted tubule really isosmotic?

It has long been known that reabsorption in the proximal convoluted tubule is isosmotic – or at least very nearly so. We consider here the possibility that there is a very small, but significant, gradient of osmotic pressure across the tubular wall, between the fluid in the lumen and the surrounding interstitial fluid.

For water to be reabsorbed by osmosis there must be an osmotic gradient somewhere, but it has been suggested that this lies within, and not across, the epithelium (the 'standing gradient hypothesis'). Such a mechanism seems to operate in some other epithelia, but, at the time of writing, its applicability to the mammalian proximal tubule is regarded as dubious.

The brochures of two kinds of commercially produced osmometer give the accuracy of measurement as 'within 2 mosmol/kg water'. One might not expect the same degree of accuracy in the context of renal micropuncture studies because of the small volumes involved, and indeed, for some of these studies, it can be said only that the osmolality of tubular fluid was within 10 mosmol/kg water of the value for systemic plasma. Would a difference of this magnitude between lumen and interstitial fluid suffice to explain the reabsorption of fluid?

Reabsorption in each millimetre length of proximal nephron has been estimated, in rats, as varying between 0.4 and 4 nl/min. Measurements of the permeability of the tubular epithelium to water (again in rats) have indicated that for each mosmol/kg water of osmolality difference, and for each 1 mm length of tubule, there would be a net volume flow (reabsorption) of about 0.4 nl/min (or possibly rather more).

5.7.1 **On the basis of the estimates in the previous paragraph, what difference in concentration between plasma and tubular fluid (the range of values) is required to explain volume reabsorption?**

5.7.2 **If the accuracy of measurements of osmolality is only ±10 mosmol/kg water, can the existence of such differences be ruled out?**

With suitable apparatus the osmolality of micropuncture samples can in fact be measured more accurately than that, e.g. with an error of only 0–1.5 mosmol/kg water. More recent studies on rat proximal tubules have shown that the luminal fluid can be more dilute than the systemic plasma by about 3 mosmol/kg water near the beginning of the tubule and by about 7.5 mosmol/kg water further along.

5.8 Work performed by the kidneys in sodium reabsorption

It has been found that the oxygen consumption of a kidney increases with the glomerular filtration rate. Because most of the filtered salt is reabsorbed, there must also be a positive correlation between the rates of oxygen consumption and of sodium reabsorption. Indeed, the relationship is essentially linear and indicates that the consumption of one oxygen molecule is associated with the transport of about 29 sodium ions. (There is also a small basal oxygen consumption which, since it persists in the absence of filtration, seems not to be related to the processes of reabsorption.)

It was calculated in Section 5.6 that a pair of typical human kidneys reabsorbs about 27 mol of sodium per day.

5.8.1 **How many mol of oxygen, as a round number, are consumed per day in transporting this amount of sodium?**

5.8.2 **What is this, expressed as l/day at standard temperature and pressure (STP)?** (1 mol of gas occupies 22.4 l at STP)

5.8.3 **What is the previous answer as a percentage of the basal oxygen consumption of the whole body if the latter happens to be 350 l/day?**

So the work of sodium reabsorption, though not huge, is certainly significant. Accordingly, it may be supposed that natural selection has acted during our evolution to keep this work to a minimum. Our remote invertebrate ancestors living in the sea almost certainly contained much higher concentrations of sodium in their extracellular fluids than we do. In typical marine invertebrates today the concentration is close to that in the ambient sea water and that concentration is about three times as high as in mammalian extracellular fluid.

5.8.4 **How much greater would the energy cost of sodium reabsorption be, for the same GFR, if our extracellular fluid contained three times as much sodium per litre as it actually does?**

Evidently there is some energetic advantage in having dilute body fluids. Nor is this the only one, for the lower the concentration of sodium in the extracellular fluid, the less work must be expended in transporting sodium out of all cells of the body in the face of continuous inward diffusion.

The work of renal sodium reabsorption could also be reduced by evolving a lower GFR, but the typical human GFR is presumably optimal for some other reason. Despite the dependence of specific metabolic rate on body size (Section 2.7), the ratio of GFR to total resting metabolic rate is much the same in large and in small mammals.

In 5.8.2 it was calculated that about 22 l of oxygen are consumed per day in reabsorbing filtered sodium. This is equivalent to 15 ml/min.

5.8.5 **Assuming that the arterial blood contains 200 ml of oxygen/l, what is the minimum blood flow, in ml/min, required to supply the oxygen for renal sodium reabsorption?**

This value is very small compared with the actual renal blood flow of 1.2–1.3 l/min. Much more sodium could therefore be filtered and reabsorbed without the need for a greater delivery of oxygen. Indeed, it is a fact often stressed that the kidneys are unusual for their high blood flow relative to oxygen consumption. Putting this differently, the ratio of GFR to renal plasma flow (i.e. the filtration fraction, usually about 0.15–0.2) is far from being limited by the oxygen requirements of sodium reabsorption.

The proximal tubules of the kidneys, where most sodium reabsorption occurs, contain large numbers of mitochondria. Since these are the site of oxidative metabolism, there ought to be a relationship between oxygen consumption and the abundance of mitochondria. Through limitations of data, we explore the matter in relation to the kidneys as a whole.

The mitochondrial content of human kidneys is not known, but in rats it has been estimated as 18.2%. Although measured in terms of volume, it should not be too inaccurate to assume this value to be 18 g/100 g. This will be used to estimate the oxygen consumption of the mitochondria in rat kidneys in ml/min per g of mitochondria. For comparison, the maximum rate in skeletal muscle is about 3–6 ml/min per g of mitochondria.

Since measurements of renal oxygen consumption in rats are not to hand, we may start with the assumptions that a pair of human kidneys utilizes 15 ml of oxygen/min in reabsorbing sodium (see above) and that they have a (typical) combined mass of 300 g.

5.8.6 **What is this rate in ml/min per g of kidney?**

This value needs to be adjusted for the difference between rats and humans. Rats, being smaller than we are, have higher specific metabolic rates than we do (see Section 2.7) and a higher value is also found for the ratio of GFR to body mass. Thus it is that the ratio of GFR to mass of kidney in rats is about 2–3 times the ratio in ourselves. Rat kidneys thus have, per g, correspondingly more filtered sodium to reabsorb.

5.8.7 **If rats use 2.6 times as much oxygen in reabsorbing sodium per g of kidney as we do (i.e. between 2 and 3 times), at what rate (in ml/min per g kidney) do their kidneys need to consume oxygen in order to power the reabsorption of sodium?**

5.8.8 **If rat kidneys contain 18 g of mitochondria per 100 g of tissue, what is the previous answer expressed in ml oxygen/min per g of mitochondria?**

Compared with the maximum rate in skeletal muscle, 3–6 ml/min per ml of mitochondria, this is quite low. In trying to explain the discrepancy, one should bear in mind that estimates have been brought together here from varied sources that may not be strictly compatible. Moreover, some allowance should be made for a 'safety margin' to cope with a higher GFR and for the fact that oxidative metabolism is required for processes other than sodium reabsorption. If, however, the general conclusion is correct, that renal mitochondria consume oxygen at a rate much lower than that of fully active muscle mitochondria, one may fairly ask why renal mitochondria should not be fewer, but correspondingly more active. I do not know the answer.

The kidneys do not use oxygen only to reabsorb sodium. Their total oxygen consumption is thus somewhat more than was calculated in question 5.8.1, probably by a litre or so per day. However, even this total does not represent the whole cost of urine production, for the heart contributes too, driving blood through the kidneys and fluid through the glomerular filters. With the body at rest, the oxygen consumption of a 300 g heart is about 24–30 ml/min (Section 3.7), or 35–43 l/day. About one-fifth of the systemic cardiac output goes to the kidneys.

5.8.9 **If the rates of oxygen consumption of the kidneys and heart are respectively 23 l/day and 40 l/day, and if the blood flow to the kidneys is**

one-fifth of the cardiac output, what is the total cost of urine production in litres of oxygen/day?

5.9 Mechanisms of renal sodium reabsorption

What can we learn about the mechanisms of renal sodium reabsorption from its cost in terms of energy or oxygen? The processes of solute transport in different parts of the nephron and collecting duct are many. In particular, sodium ions may pass through the apical (luminal) cell membranes through pores and on carriers, alone or in association with other solutes (accompanying glucose, exchanging with hydrogen ions, etc.). Sodium may also pass through the basolateral membranes of the cells in various ways, including by electrogenic cotransport with bicarbonate, but the predominant mechanism there is that of the sodium pump (Fig. 5.4). In other words, most of the sodium passing through the cells is actively transported through the basolateral membranes in exchange for potassium and at a cost of one ATP molecule for each three sodium ions transported. How does this square with the idea (Section 5.8) that about 29 sodium ions are associated with the consumption of one oxygen molecule?

To answer this, we need to know the relationship between oxygen consumption and ATP production, but this relationship depends on the nature of the respiratory substrate. The renal medulla utilizes mainly glucose, but in the cortex, where most of the sodium reabsorption occurs, the main substrates are fatty acids. When the substrate is a fatty acid, the ratio of ATP molecules produced per oxygen molecule depends on which fatty acid is involved, but we may take palmitate as representative. This has an ATP/O_2 ratio of 5.6 (as compared with 6.3 for glucose).

5.9.1 **Assuming that 29 sodium ions are associated with the consumption of one oxygen molecule and that the ATP/O_2 ratio is 5.6, what is the ratio of sodium ions reabsorbed to ATP molecules synthesized and broken down?**

As a round number, the answer is the same for palmitate and glucose and it is well above the ratio of 3 that typically corresponds to transport by Na, K-ATPase. This may be surprising inasmuch as it is easy to see how the ratio could be less than 3 – through back-leakage of Na and through consumption of ATP in other ways. The discrepancy may be accounted for in terms of two processes that do not directly involve Na, K-ATPase.

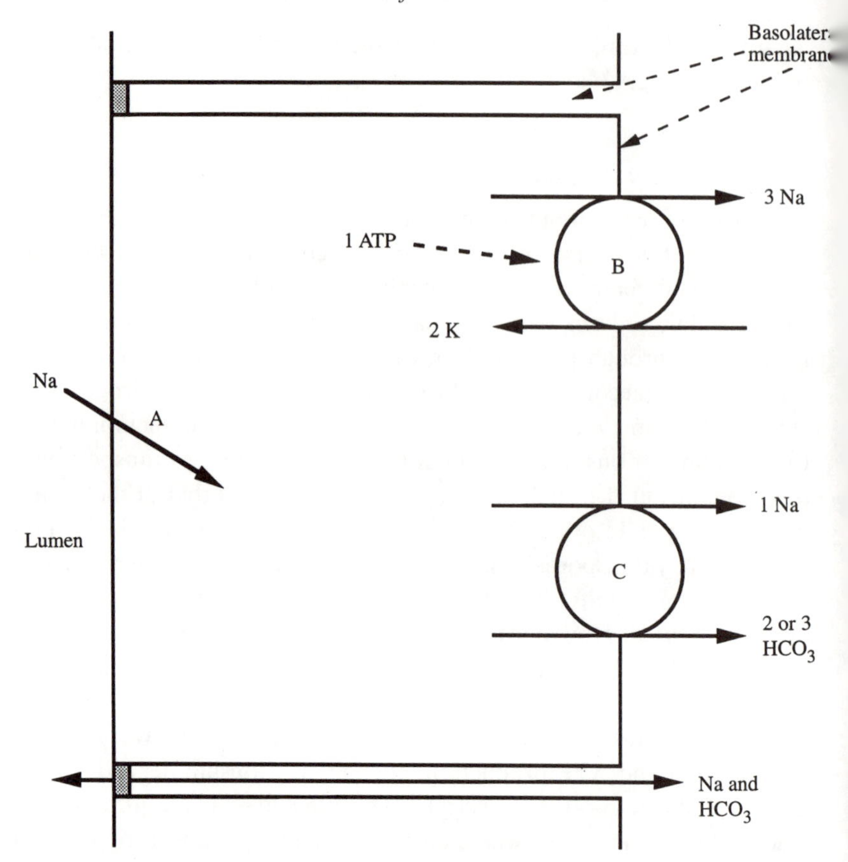

Fig. 5.4. Some mechanisms of sodium reabsorption in the proximal tubule. From the lumen, to the left, sodium enters the cell in various ways down its electrochemical gradient (A). It leaves through the basolateral membrane via the sodium pump (B) and by electrogenic cotransport with bicarbonate (C). ('Cotransport' signifies that the sodium and bicarbonate are transported together, in the same direction and on the same carrier in the cell membrane. The cotransport is called electrogenic because more bicarbonate ions are transported this way than sodium ions, so that there is a net transfer of charge.)

One of these is the passage of sodium between the cells, i.e. through the paracellular pathway. This is distinctly leaky to ions in the proximal tubule and it is an important route for sodium reabsorption in the loop of Henle. The other process is one that has been mentioned already, namely the passage of sodium through the basolateral membranes together with bicarbonate. The proportion of sodium reabsorbed in this way is quite small, as will be shown next.

We now calculate an upper limit for the proportion of sodium that might be reabsorbed by the cotransport of sodium and bicarbonate through the basolateral membranes of the epithelial cells. As already noted (Section 5.6), sodium and bicarbonate are reabsorbed by the whole kidney in roughly the same ratio as they occur in plasma. This ratio (i.e. Na/HCO_3) is about 6. Basolateral electrogenic cotransport of the two ions involves the movement of either two or, more probably, three bicarbonate ions with each sodium ion. (At the time of writing, it is not clear which ratio is correct.)

5.9.2 **If *all* the bicarbonate reabsorbed in the kidneys were to cross the basolateral membranes of the epithelial cells of the tubules in association with sodium, what percentage of the total reabsorbed sodium would have to be reabsorbed with bicarbonate (and therefore not by Na, K-ATPase)?** Assume either ratio ($2HCO_3/1Na$ or $3HCO_3/1Na$); the first yields the higher percentage.

The percentage of sodium reabsorbed this way must be even lower if some bicarbonate passes through the basolateral membranes unaccompanied by sodium.

5.10 Autoregulation of glomerular filtration rate, glomerulotubular balance

The rate at which urine is produced is necessarily equal to the glomerular filtration rate (GFR) minus the net rate of reabsorption of fluid from the renal tubules. Thus, a rate of urine production of 1 ml/min might correspond to a GFR of 125 ml/min and a rate of reabsorption of 124 ml/min, i.e.

$$1 \text{ ml urine/min} = 125 \text{ ml/min (filtered)} - 124 \text{ ml/min (reabsorbed)}.$$

5.10.1 **What is the rate of urine flow if (a) the GFR rises to 126 ml/min without change in the rate of reabsorption, and (b) the GFR stays at 125 ml/min and the rate of reabsorption falls to 123 ml/min?**

Since urine flow changes so much in the stated circumstances, it is obvious that there must be mechanisms controlling both the GFR and the rate of fluid reabsorption. (Note that, although one may postulate rates like those above, the rate of reabsorption cannot be measured directly and the change in GFR is too small to be detected experimentally.)

Two phenomena have been revealed experimentally that correspond to the two requirements for control thus identified. One is the autoregulation of GFR, whereby GFR remains nearly constant in the face of variations in the blood pressure in the renal arteries. (Good autoregulation occurs over only a limited range of arterial blood pressure and other mechanisms can override it, so that, for example, the GFR may fall during exercise. Autoregulation of GFR is linked to autoregulation of renal blood flow.) The second phenomenon is known as glomerulotubular balance and consists of the adjustment of the rate of fluid reabsorption to match, nearly, such changes as do occur in GFR. The difference between GFR and rate of fluid reabsorption (i.e. the rate of urine flow) is thus far less variable than might otherwise be. Neither autoregulation nor glomerulotubular balance is total, and this fact is essential to the line of thought in Section 5.11.

5.11 Renal regulation of extracellular fluid volume and blood pressure

Many mechanisms contribute to the regulation of extracellular fluid volume, involving hormones (angiotensin, aldosterone, atrial natriuretic hormone, vasopressin), renal reflexes and the colloid osmotic pressure of the blood plasma. The mechanism that concerns us here is the direct effect of arterial blood pressure on the rate of urine production. Some people regard the mechanism in question as the most important determinant of blood pressure in the long term.

Again we consider the necessary dependence of urine flow on GFR and rate of fluid reabsorption (Section 5.10). As a starting condition, suppose that the mean arterial blood pressure of an individual is 100 mmHg, that the GFR is 125 ml/min, and that urine is being formed at 1 ml/min.

5.11.1 **What is the rate of fluid reabsorption?**

Now suppose that the blood pressure rises from 100 to 140 mmHg. (This rise could be due to the accumulation of excess extracellular fluid, with some of this being in the blood.) Bearing in mind the phenomenon of autoregulation of GFR, and remembering that this is not completely effective, let us suppose that the GFR rises, not in proportion to blood pressure, but only to 135 ml/min (a rise of 8%, rather than of 40%). Finally, let us postulate that fluid reabsorption also rises (glomerulotubular balance), by about 6% rather than the full 8%, to 131 ml/min. The only values needed for the next calculation are these new rates of filtration and reabsorption and the original rate of urine flow (1 ml/min).

5.11.2 **By what factor does the rate of urine production change in response to the rise in blood pressure?**

It has been known since 1843 that urine flow increases with perfusion pressure – in isolated kidneys and *in vivo*. The phenomenon is called 'pressure diuresis'. It is accompanied by increased excretion of sodium ('pressure natriuresis'). The above calculations suggest that pressure diuresis is almost inevitable, but is a small rise in GFR really responsible? If so, that may be only part of the story, for pressure diuresis is thought to occur in the absence of any detectable rise in GFR. A more important mechanism may be a reduction in fluid reabsorption in the papillae as a result of increased pressure in the vasa recta and surrounding interstitial fluid.

5.12 Daily output of solute in urine

The total amount of solute excreted daily in the urine expressed in milliosmoles (the 'excreted solute load') is a valuable concept in the clinical interpretation of water imbalance. Representative values for the daily urinary solute output are required in following sections.

The main solutes in urine are urea and inorganic salts. Table 5.1 shows typical excretion rates for the substances most generally abundant in the urine. Actual ranges depend very much on diet and are by no means as well defined as is implied by Table 5.1. Thus, to take the case of sodium, intake and excretion may be as high as 500–600 mmol/day in parts of Japan and they may be as low as 2–10 mmol/day in the Amazon jungle and highland New Guinea, where plants and soil are very low in salt. The rate of urea excretion increases with the protein content of the diet, but when the latter is zero the rate of urea excretion falls only to about 100 mmol/day. This is because of the breakdown of body protein.

An estimate of the typical range of solute output may be obtained by summing all the minimum values in Table 5.1 and also all the maximum values. The results will suffice for present purposes, but the method has some deficiencies that should be noted. It could give too great a spread inasmuch as the two extreme values would require that every solute be simultaneously minimum or maximum. On the other hand, the ranges for the individual solutes are not themselves extremes. Because the units used

Table 5.1 *Typical renal excretion rates on a Western diet*

Substance	Excretion rate (mmol/day)
Urea	270–580
Sodium	80–200
Chloride	80–200
Potassium	40–150
Sulphate	18–28
Ammonium	30–70

in Table 5.1 are mmol/day instead of mosmol/day as required, the totals should be reduced by roughly 10% to allow for non-ideal behaviour of the urine in terms of its physical chemistry. However, an arbitrary 10% can be added back to cover the variety of untabulated solutes. In brief, it should suffice to sum the values given so long as the estimates are regarded as approximations.

5.12.1 **On the basis of the data in Table 5.1, what seems to be a likely range for the solute output in mosmol/day?**

Daily solute output may also be obtained directly from measurements of osmotic concentrations and volumes of pooled urine samples.

Ammonium ions and urea (see entries in Table 5.1) both contain nitrogen and this may now be related to the dietary protein from which most of it ultimately derives. Since each molecule of urea contains two nitrogen atoms, the rate of nitrogen excretion (ignoring untabulated nitrogen compounds) is, according to the values in Table 5.1, about 570–1230 mmol/day, or 8–17 g/day. The number of grams of protein containing this amount of nitrogen is readily calculated, given the fact that 1 g of nitrogen is equivalent to 6.25 g of protein.

5.12.2 **To how much protein does this range of daily nitrogen excretion correspond?**

Daily allowances for protein, as recommended by the Department of Health and Social Security for the diets of young men and women, are in this range.

5.13 The flow and concentration of urine

Section 5.12 introduced the notion of daily solute output or excreted solute load. Values on a Western diet are generally in the range of 500–1200 mosmol/day. The daily solute output is equal to the product of daily urine volume and average solute concentration. Alternatively,

$$\text{urine concentration in mosmol/l} = \frac{\text{solute output in mosmol/day}}{\text{urine flow in l/day}}. \quad (5.5)$$

Although formulated in terms of the whole day, the equation may of course be applied to shorter time scales. (Osmotic concentrations are given elsewhere as 'mosmol/kg water' as that is usually most appropriate, but here it is more convenient to work in terms of volumes.)

As long as the rate of solute output remains constant, concentration varies inversely with flow rate. The next three questions illustrate this in the context of rates and concentrations found for normal humans.

5.13.1 **An individual produces isosmotic urine so that its concentration, like that of the plasma, is about 300 mosmol/l. If the solute output is 750 msmol/day, what is the rate of urine production?**

The rate of urine production is usually near half this answer.

5.13.2 **So, for the same solute output, but half the rate of urine production, what would the (more typical) concentration be?**

5.13.3 **For the same solute output, what concentrations would correspond to the two moderately extreme flow rates of (a) 15 l/day and (b) 0.6 l/day?**

The unusually high flow rate of 15 l/day could be associated with diabetes insipidus or the excessive water drinking that occurs in some psychiatric disorders. The same rate (≡ 10 ml/min) is commonplace after a single large drink. For a complete curve of concentration against flow rate, see Notes and Answers. Note that this curve applies only to the chosen solute output of 750 mosmol/day; moment-to-moment variations in an individual would only follow that curve exactly if the solute output were to stay constant at that value. (Yet Fig. 5.3 implies an increase in the urea component with increasing urine flow.)

The maximum concentration of human urine is actually nearer to 1400 mosmol/l than to the concentration just calculated, but our ability to concentrate urine depends on the solute output. Counter to intuition perhaps, the more solute is excreted, the lower is the maximum concentration. Further calculations reveal why this might be.

While in the proximal tubules, and again in the distal tubules during antidiuresis, the tubular fluid is isosmotic to plasma. Before leaving the kidneys as concentrated urine, fluid must become concentrated from that initial 300 mosmol/l by the reabsorption of water in the collecting ducts. The rate of water reabsorption may be estimated from data on urine such as those used above.

5.13.4 **Suppose that the solute output is 600 mosmol/day and that the urine concentration is 1200 mosmol/l. The urine flow rate is thus 0.5 l/day. While still isosmotic with plasma at 300 mosmol/l, the volume of fluid necessary to contain that solute output (still within the renal cortex) must be 600/300 = 2 l/day. What is the rate of subsequent fluid reabsorption?** (Ignore here and below the fact that there is additional solute reabsorbed – with water – in the collecting ducts; it does not affect the ultimate conclusion.)

5.13.5 **On the basis of similar calculations, what would the equivalent rate of reabsorption have to be if the solute output were doubled to 1200 mosmol/day and if the final concentration of the urine were again 1200 mosmol/l?**

With so much extra water to be reabsorbed into the medullary interstitium it is not surprising that the concentrating ability of the kidneys should be reduced. With the doubled solute output (1200 mosmol/day), the maximum concentration would in fact fall to about 1000 mosmol/l.

5.13.6 **What would then be the rate of urine production?**

Comparing this rate with the rate of 0.5 l/day given for question 5.13.4, we see that under conditions of maximum antidiuresis an increase in solute output leads to an increase in the flow of urine. In more extreme cases the phenomenon is called 'osmotic diuresis'. It is strikingly shown when glucose is excreted in diabetes mellitus (Section 5.4). Sucrose and mannitol can be administered as osmotic diuretics, these being, like inulin, filtered from the plasma, but not reabsorbed.

Following the same line of argument, it is evident that a person short of water (in the desert, or adrift in a lifeboat) should avoid increasing the excreted solute load. From this point of view, it is better to eat carbohydrate than protein, since the latter yields urea. (Note, however, that some urea is required for the functioning of the medullary countercurrent mechanism.) It is well known that one should not drink sea water in this situation because of its high salt content.

5.13.7 **Referring back to questions 5.13.5 and 5.13.6, let us suppose that the extra solute output of 600 mosmol/day in question 5.13.6 was due to the salts of sea water, incautiously drunk by the individual in question. The concentration of sea water is about 1000 mosmol/l, so that the extra solute corresponds to 600 ml of it. We have already calculated that the rate of urine production would increase from 0.5 l/day to about 1.2 l/day. Would the extra loss of water in urine be more or less than the volume of sea water ingested, and by how much?**

The house mouse and numerous desert rodents are much better adapted to conserving water than we are. The Australian hopping mouse (*Notomys*) can concentrate its urine to 9000 mosmol/l.

5.13.8 **Consider again an individual like that in question 5.13.4, producing 0.5 l of urine per day at 1200 mosmol/l, with a daily solute output of 600 mosmol. If the urine could be concentrated to 9000 mosmol/l, what would be the daily saving in water?**

Would such a saving be worthwhile in a human?

We turn now to the extremes of diuresis. What are the highest rates of urine flow and lowest concentrations of urine? It can be hard to find exact values for these. One reason is that the extremes must depend on circumstances, including solute output, but it is also true that most textbooks give much more attention to antidiuresis. Maximum urine volumes in a day are sometimes to be found, given for example as '15–20 l' or '20–25 l'. Minimum concentrations may be given as '30 mosmol/l or less', 'about 50 mosmol/l' or '50–75 mosmol'. What does this inconsistency mean – that the matter has not been adequately explored, or that the minimum concentration depends on circumstances? Short of seeking original data or other accounts, we can hardly resolve that, but we can try putting pairs of these values together to see whether the implied outputs of solute are likely. For this purpose, let us take the highest flow rate (25 l/day) and the low concentrations of 30 and 50 mosmol/l. It seems reasonable that the extremes of flow and dilution should go together, although it should be kept in mind that the values do come from different sources.

5.13.9 **How much solute is excreted per day, in mosmol, when urine flows at 25 l/day and its concentration averages (a) 30 mosmol/l and (b) 50 mosmol/l?**

In question 5.12.1, typical values for solute output were estimated as about 500–1200 mosmol/day. Of the values just calculated, the first is within that range. The second, though possible, is too high to be typical. What we may conclude from these results is that two values may be coupled in our minds as extremes without obvious inconsistency – a maximum flow rate of about 25 l/day (=17 ml/min) and a minimum concentration of about 30 mosmol/l. (As a matter of fact, this pairing of rate and concentration has been recorded.)

It should not be concluded that these two extreme values must apply to all water-loaded individuals under all circumstances, or even that they are commonly achieved. Indeed, the minimum dilution of the urine following considerable water loading is often above 50 mosmol/l. The extremes are more likely to be sustainable over minutes than over whole days.

5.14 Beer drinker's hyponatraemia

The fact that there is a maximum dilution to the urine (see the discussion at the end of Section 5.13) has important implications for those individuals who drink large amounts of fluid, but who have only a small solute load to excrete. Beer has a low salt content and its dedicated drinkers often tend to forgo the normal sources of nutrition and so have little urea, salt and other solute to excrete. This both limits the excretion of water and leads, by a general dilution of the body fluids, to hyponatraemia (low plasma sodium). After subsisting for a time on beer, patients are sometimes admitted to hospital suffering from debility and dizziness, and with plasma sodium even below 110 mmol/l. To explore how this happens, we return to equation (5.5):

$$\text{urine concentration in mosmol/l} = \frac{\text{solute output in mosmol/day}}{\text{urine flow in l/day}}. \quad (5.5)$$

The normal urinary output of solute is about 600–1200 mosmol/day (Section 5.12); let us assume here that in a particular beer drinker it is only 240 mosmol/day. Let us also suppose that he has 6 l of water to dispose of daily as urine (as compared with a more typical 1–2 l/day). The average urine concentration through the day may now be calculated.

5.14.1 **If a beer drinker excretes solute at a rate of 240 mosmol/day with a urine flow of 6 l/day, what is the average concentration of the urine in mosmol/l?**

Even though the rate of urine flow is well below the normal maximal rate of 20–25 l/day, the calculated concentration is very low (Section 5.13). To understand how water retention would occur in such an individual, one has only to postulate that urinary dilution is not actually as effective as this over the whole day. If the mean concentration were, say, 60 mosmol/l, the volume of urine produced in a day would be 240/60 = 4 l, leaving 2 l unexcreted. Actual measurements show that the concentration of the urine on admission to hospital can exceed 60 mosmol/l.

An obvious remedy for the water retention is to increase the intake of salt and hence the amount of solute available for excretion. Salted crisps would obviously help, but some drinkers salt their beer.

Compulsive water drinking, as is seen in some psychotic patients, also leads to hyponatraemia. Associated with the hyponatraemia there is often a significant impairment of urinary dilution.

5.15 The medullary countercurrent mechanism – applying the principle of mass balance

Fluids flow into and out of the renal medulla within the vasa recta, loops of Henle and collecting ducts. The total amount of fluid or water entering the medulla must equal the total amount leaving, and this is true of the solutes also. This obvious point is sometimes called the principle of mass balance. (It was applied in relation to clearance, i.e. equation (5.3).) Many accounts of the medullary countercurrent mechanism have little to say on this and leave some readers uneasy on the matter. It was for such people that the following calculations were originally intended, but it happens that other points of interest also emerge.

This Section constitutes the longest sustained argument in the book. Partly to avoid complicating it further, a minor simplification is adopted that affects the values rather little and the conclusions not at all. In the context of osmotic concentrations the fact that whole blood contains only about 80% water is ignored. The units used are mosmol/l rather than mosmol/kg water (Section 6.9).

As a preliminary to focussing on the medulla, it may help to apply the principal of mass balance to the simpler case of the whole kidneys. Suppose that they are producing urine, in antidiuresis, at a rate of 0.5 ml/min. The renal blood flow is a typical 1200 ml/min, say, and the osmotic concentrations of the afferent blood and the urine are 300.0 and 1200 mosmol/l respectively. How much would one expect the blood to be diluted as it passes through the kidneys? The amount of solute entering the kidneys in the blood and leaving in urine are respectively $300 \times 1200/1000 = 360$ mosmol/min and $1200 \times 0.5/1000 = 0.6$ mosmol/min. The flow of (slightly hyperosmotic) lymph from the kidneys is slow enough not to be significant in this context. According to these values, 359.4 milliosmol leave the kidneys each minute in 1199.5 ml of blood.

5.15.1 **What, in mosmol/l to one decimal place, is the concentration of the blood leaving the kidneys?**

Compare this with the concentration in the afferent blood.

Fig. 5.5 is a reminder of fluid and solute movements relevant to the countercurrent mechanism. Some of the details do not concern us here, such as the otherwise important distinctions between salt and urea and between long and short loops of Henle. Even the countercurrent flows themselves, of blood and tubular fluid, will not be considered. Especially important here is the fact that dilute fluid flows out of the loops of Henle to the distal tubules, leaving excess solute in the medulla. It is this that makes possible the flow of concentrated fluid (i.e. urine) out of the medulla. In other words, if a concentrated fluid leaves, then so must a dilute one.

Fig. 5.6 shows only the fluids entering and leaving the medulla. Solute concentrations are given for most of these, and symbols representing flow rates (for the pair of kidneys) are also shown. The transition between

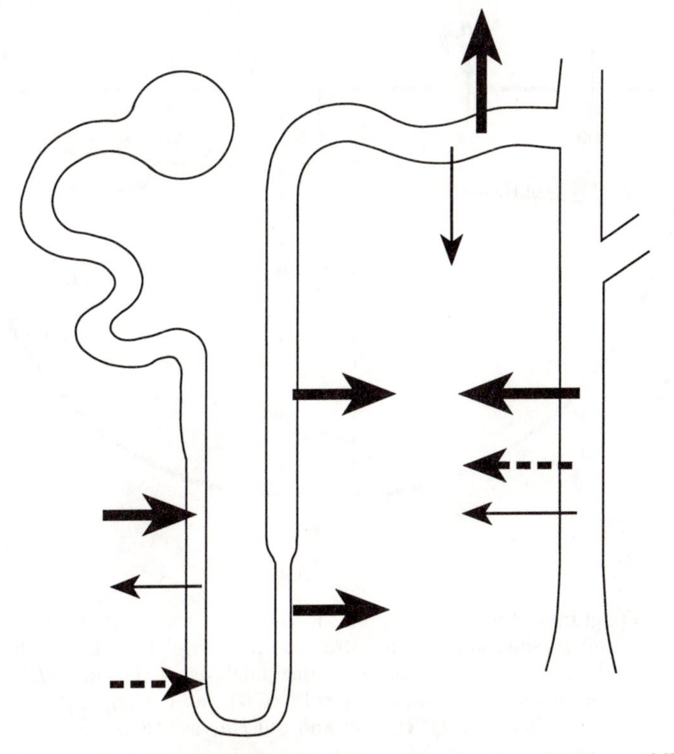

Fig. 5.5. Diagram of a long-looped nephron and collecting duct in antidiuresis to show transepithelial movements of water, sodium and urea that are relevant to the countercurrent mechanism. →: sodium salts. →: urea. --→: water.

cortex and medulla is defined here, not anatomically, but as the limit of the medullary osmotic gradient.

Applying the principle of mass balance only to fluid flow, we have (using the conventions of Fig. 5.6):

$$(VR)_{in} + (LH)_{in} + (CD)_{in} = (VR)_{out} + (LH)_{out} + (CD)_{out}. \qquad (5.6)$$

As to the numerical values applicable in antidiuresis, the two best known are the rate of flow from the collecting ducts, $(CD)_{out}$, since this is the rate of urine production, and the rate of flow into the loops of Henle. The rate of urine production will be taken as 0.5 ml/min. The rate of flow into the loops of Henle can be taken as the glomerular filtration rate, say 120 ml/min, less 67–80%, these being textbook values for the percentage reabsorption of fluid in the proximal convoluted tubules. From within the resulting range of 24–40 ml/min let us choose for $(LH)_{in}$ a round value of 30 ml/min.

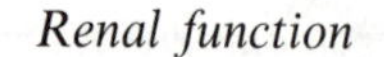

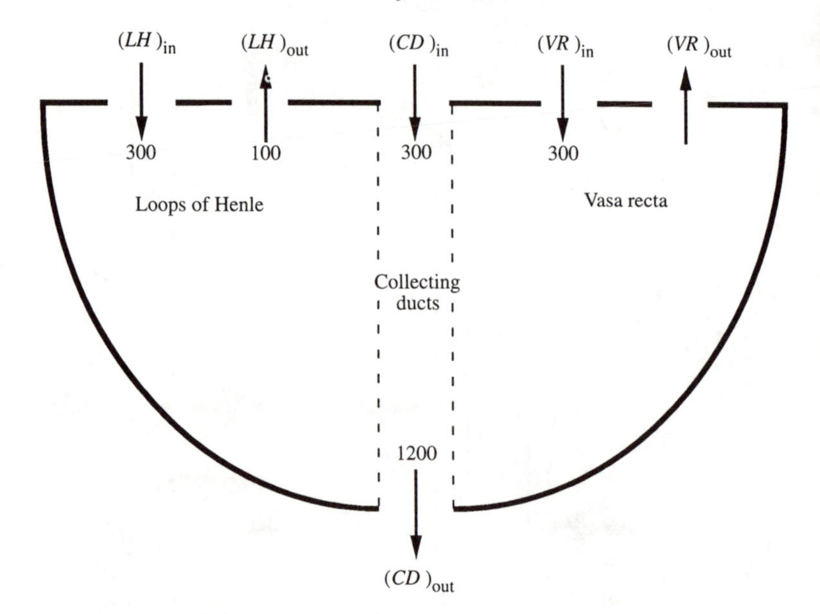

Fig. 5.6. Diagram of the renal medulla showing the inward and outward flows of fluid. The numbers show solute concentrations in mosmol/l and the symbols stand for rates of flow as used in the mass-balance analysis: $(LH)_{in}$ and $(LH)_{out}$, into and out of the medulla *via* loops of Henle; $(CD)_{in}$ and $(CD)_{out}$, in and out *via* collecting ducts; $(VR)_{in}$ and $(VR)_{out}$, in and out *via* vasa recta.

Now for $(CD)_{in}$. In the circumstances of question 5.13.4, $(CD)_{in}/(CD)_{out}$ was at least 4, the factor by which the tubular fluid was concentrated by reabsorption of water (i.e. 1200/300). Recognizing that some urea and salt are reabsorbed in the medullary collecting ducts, and so, also, is additional water, let us take the value of $(CD)_{in}$ somewhat arbitrarily, and knowing no better, as 4 ml/min, rather than 4×0.5 ml/min.

Similar reasoning may be applied to the loops of Henle on the assumption that their contents are concentrated in the descending limb mainly by osmotic reabsorption of water (which is true in the rabbit, but less so in the rat) and that no water is subsequently gained or lost in the ascending limb (the low permeability of which to water is so important to the countercurrent mechanism). In those long loops extending all the way to the tip of the papilla, there would be the same fourfold increase in concentration as in the collecting ducts. If all loops extended to the tip of the papilla, then $(LH)_{in}/(LH)_{out}$ would equal 4 and $(LH)_{out}$ would be $30/4 = 7.5$ ml/min. However, the ratio must be lower than 4 for the greater number of shorter loops. Accordingly, a more realistic average value for $(LH)_{out}$ must be somewhere between 7.5 and 30 ml/min. Let us say 20 ml/min.

For most of these quantities, true representative values for human kidneys in antidiuresis are not known, but those used here do not seem to conflict too much with experimental evidence from smaller mammals. To summarize, the chosen values are as follows:

$$(LH)_{in} = 30 \text{ ml/min} \qquad (LH)_{out} = 20 \text{ ml/min}$$

$$(CD)_{in} = 4 \text{ ml/min} \qquad (CD)_{out} = 0.5 \text{ ml/min}.$$

5.15.2 **On the basis of equation (5.6) and these values, what is $[(VR)_{out} - (VR)_{in}]$?**

One may well be uneasy about the accuracy of the answer, but its sign must be correct. In other words, more blood must leave the medulla than enters it, having taken up excess water in the vasa recta. This much is evident from equation (5.6), given simply that $(LH)_{in}$ exceeds $(LH)_{out}$ and that $(CD)_{in}$ exceeds $(CD)_{out}$.

There is a problem in comparing $[(VR)_{out} - (VR)_{in}]$ with either of its components, for medullary blood flow is not easily measured. Reflecting the variety of measurements made on experimental animals, textbooks give values for medullary blood flow that range from '0.7%' to '10–15%' of the total renal blood flow, i.e. 8–200 ml/min. At least one can say that $[(VR)_{out} - (VR)_{in}]$ calculated in question 5.15.2 is not insubstantial by comparison with any such values!

Considering now the movements of solute and referring again to Fig. 5.6, these may be calculated as products of flow rates and concentrations. All three fluids entering the medulla do so at the same concentration, taken here as 300 mosmol/l. For the urine in antidiuresis, the value of 1200 mosmol/l may be used as before, while another value common in textbooks, namely 100 mosmol/l, suffices for the dilute fluid leaving the loops of Henle. This leaves one unknown concentration, that of the blood leaving the medulla, denoted B. Equating the amounts of solute entering and leaving the medulla (in $n$osmol/min), we have:

$$300\{(VR)_{in} + (LH)_{in} + (CD)_{in}\} = B(VR)_{out} + 100(LH)_{out} + 1200(CD)_{out}. \tag{5.7}$$

Rather than trying to use this equation directly, let us combine it with equation (5.6) in such a way as to remove three of the flow rates:

$$(VR)_{out}\{B - 300\} = (LH)_{out}\{300\text{–}100\} + (CD)_{out} \cdot \{300 - 1200\},$$

or

$$B = 300 + \{200(LH)_{out} - 900(CD)_{out}\}/(VR)_{out}. \qquad (5.8)$$

Unfortunately we are left with that very uncertain quantity, $(VR)_{out}$. Let us try out two extreme values.

5.15.3 **Taking the value of $(VR)_{out}$ as (a) 30 ml/min and (b) 200 ml/min, and other flow rates as suggested above, what is the value of *B*, the concentration of the blood leaving the vasa recta, in mosmol/l?**

Regardless of the actual medullary blood flow, it is clear that the blood becomes substantially concentrated and that the vasa recta carry away substantial amounts of excess solute as well as excess water.

Solution (a) in 5.15.3 may well seem intuitively to be an unlikely one, for the increase in blood concentration of 118 mosmol/l seems high, and so does the ratio of $(VR)_{out}$ to $(VR)_{in}$, which is 1.8. (To check the latter, remember that $[(VR)_{out} - (VR)_{in}]$ was calculated in 5.15.2 as 13.5 ml/min.) One may back one's intuition with further calculation, using again the principle of mass balance. This we do here in terms of a simplified model of a single representative capillary loop (Fig. 5.7). Here is the situation to be considered in relation to solution (a) in 5.15.3:

Blood leaves the medulla at a rate of 30 ml/min, with a concentration of 418 mosmol/l. It enters the medulla at a rate of (30 – 13.5) = 16.5 ml/min with a concentration of 300 mosmol/l. Within the medulla the blood becomes concentrated maximally to a concentration of *C* mosmol/l. This occurs by diffusion of water out of the blood (at a rate of *W* ml/min) and of solutes into it, while the subsequent process of dilution involves the diffusion of water into the blood and of solutes out of it (at a rate of *S* nmol/min).

The principle of mass balance may now be applied to the solutes at the top and bottom of the ascending limb of the capillary loop. The rates at which solutes pass these points are calculable (in nosmol/min) as the products of the concentrations and the rates of blood flow. Thus,

$$418 \times 30 = C(16.5 - W) - S. \qquad (5.9)$$

Therefore,

$$C = (12\,540 + S)/(16.5 - W) \qquad (5.10)$$

The values of *S* and *W* are not known, but at least for the moment let

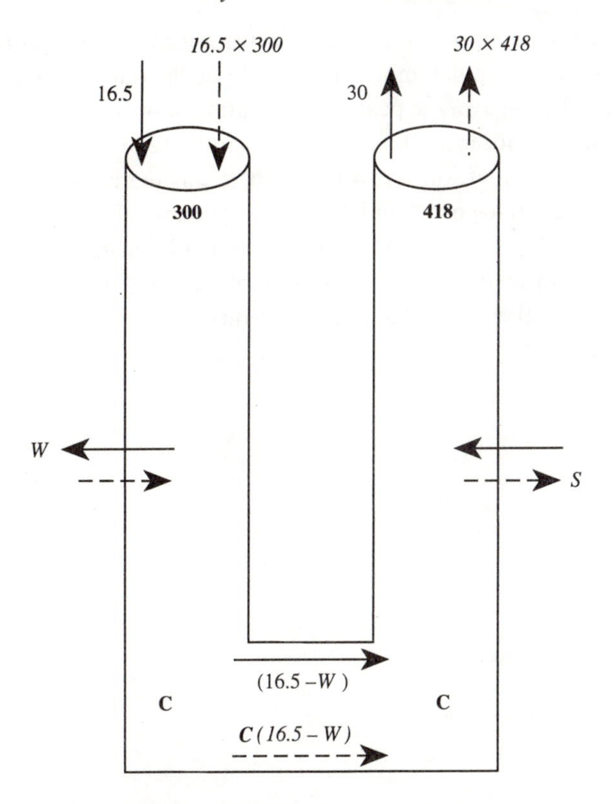

Fig. 5.7. Water and solute flows along and through the walls of the vasa recta. In this dubious model all the vasa recta are represented by a single capillary loop. Bold numbers show osmolarities of the blood (mosmol/l), with **C** being the maximum osmolarity at the bend of the loop. *W* is the rate at which water leaves through the endothelium of the descending limb. *S* is the rate at which solutes leave the ascending limb. Solid arrows, movements of water (rates shown in ml/min); dashed arrows, movements of solute (rates shown in nosmol/min, in italics).

us try giving them both the convenient but improbable value of zero. (A better justification for doing this will be apparent later!)

5.15.4 **According to equation (5.10), what then is the value of *C*, the concentration of the blood when maximally concentrated within the medulla?**

This answer needs to be assessed in relation to the real medulla, in which some blood becomes concentrated to the chosen value of 1200 mosmol/l, but in which most of the blood, not penetrating so deeply, is concentrated much less. Because much more blood perfuses the outer medulla than the inner medulla, the average peak concentration, corresponding to C in the model, must be closer to 300 mosmol/l than to 1200 mosmol/l. Not only is the answer to question 5.15.4 too high, it was made as low as possible by taking the values of W and S as zero. The model, with $(VR)_{out}$ equal to 30 ml/min, is therefore not credible and this means that the lowest values given in textbooks for medullary blood flow are probably wrong. (To proceed further in print and try to predict medullary blood flow would be foolhardy!)

6

Body fluids

Renal function and acid–base balance are treated under separate headings. In this chapter the first questions concern effects of ingesting water and potassium, and the movements of ions between extracellular fluid and tissues – cells in Section 6.3 and bone mineral in Section 6.4.

Section 6.5 on the important principle of electroneutrality then leads us more in the direction of physical chemistry – to Donnan equilibria and colloid osmotic pressure (Sections 6.6 and 6.7). Proteins may influence ionic concentrations not only through the Donnan effect, but, when the concentrations are expressed in molar rather than molal terms, by simple dilution; the distinctions between molarity and molality and between osmolarity and osmolality are explored in Sections 6.8 and 6.9, with a view to making the distinctions less tiresome for those who need to acknowledge them.

Sections 6.10 and 6.11 are about things more commonly found under such headings as 'excitable tissues' or 'nerve and muscle' (Chapter 8), these being the relationships amongst membrane potentials and trans-membrane gradients of sodium and potassium.

Some aspects of water balance are discussed elsewhere – metabolic water (Section 2.12), water loss in expired air (Section 4.6) and urine (Chapter 5). Osmoles and osmotic pressures are treated briefly in Notes and Answers.

6.1 Another drink of water

The sensitivity of osmoreceptors

As noted earlier, the brochures for at least two kinds of commercial osmometer give their accuracy as 'within 2 mosmol/kg water'. How do our hypothalamic osmoreceptors compare?

6.1.1 **The body of a man contains 49.5 l of water. He drinks 0.5 l of water. Suppose that all this water is absorbed, at first with negligible excretion. What are (a) the percentage decrease in osmotic concentration of the body fluids and (b) (given an initial concentration of 300 mosmol/kg water) the absolute decrease?**

Half a litre of water is usually enough to produce a small diuresis in someone who is in a normal (neutral) state of water balance and the typical diuretic response suggests that our hypothalamic osmoreceptors perform at least as well as the above osmometers. Indeed, a mere 1% fall in osmotic concentration is enough to effect a significant reduction in the secretion of antidiuretic hormone. Note, however, that the argument is not watertight as it stands, for the arterial blood supplying the osmoreceptors, and even more so the blood leaving the gastrointestinal tract, would temporarily be more diluted than the other body fluids. For a full analysis of the situation, the time courses of gastrointestinal absorption, mixing and diuresis should also be taken into account.

6.2 Cells as 'buffers' of extracellular potassium

Food is derived largely from plant and animal cells, and cells contain much potassium. Let us consider the effects of ingesting a specific amount of potassium, namely 30 mmol. This could be contained in, say, 300 g of skeletal muscle, hot-pot or banana, or 200 g of raw potatoes.

6.2.1 **Suppose that these 30 mmol of potassium were to be absorbed into 15 l of extracellular fluid without any of it being excreted or entering other body compartments. By how much would the concentration rise? What would the concentration then be if it were previously 4.5 mmol/l?**

Ventricular fibrillation may occur when the level of plasma potassium exceeds about 6 mmol/l – although clearly not when it happens, as it often does, as the result of exercise! Considering that a person might not only absorb the above amount of ingested potassium, but also take exercise, the initial assumption that all the potassium stays in the extracellular fluid will not do.

An obvious thought is that the excess potassium is as rapidly excreted by the kidneys as it is absorbed from the gut, but it is also a possibility that some of the potassium temporarily enters the cells. In fact the latter is an important mechanism of potassium homeostasis and although our calculation does not suffice to prove it, there is plenty of experimental evidence that the cells in general do act to 'buffer' extracellular potassium. Hormones such as insulin and catecholamines are involved in this mechanism.

Experimentally, subjects have taken seven or more times the above 30 mmol of potassium during the course of a day, with plasma potassium increasing by less than 0.3 mmol/l.

At this point it would be satisfying to calculate the rate at which the kidneys might excrete the ingested potassium and relate that to the rate of absorption from the gastrointestinal tract. Unfortunately the latter has not been well studied. It does seem, however, that absorption from the gut would take long enough (a few hours) for renal excretion to be proceeding at a significant rate.

In considering the role of the cells in buffering extracellular potassium, it helps to be aware of the distribution of potassium within the body. As round numbers, the concentration of potassium in the cells averages 150 mmol/l water and the concentration in the extracellular fluid is 5 mmol/l water. The volumes of intracellular and extracellular water are respectively about 30 and 15 l.

6.2.2 **On the basis of those values, how much potassium is there in (a) all the cells of the body and (b) the extracellular fluid?**

6.2.3 **Assuming initial conditions as just specified, if all the above 30 mmol of ingested potassium were to enter the cells, what would be the percentage increase in intracellular concentration?**

Would such a change be experimentally detectable? Is it likely that the change would influence cell function significantly?

6.3 Assessing movements of sodium between body compartments – a practical difficulty

The problem discussed here is relevant mainly to experimental studies, but is more generally worth pursuing as an exercise in the application of both elementary ideas and useful quantities.

Suppose that there is net movement of a substance between tissues and extracellular fluid within the whole body and that this movement needs to be quantified experimentally. Often it suffices to note how the concentration changes in the blood plasma. This is the case with, say, potassium or glucose, but it is not, as will be seen, the case with sodium.

To consider a particular example, suppose that sodium and potassium move with bicarbonate from cells to extracellular fluid and that they do so in a ratio of three sodium ions to one potassium ion and four bicarbonate ions. This situation and these ratios have been chosen because similar ionic shifts are thought to occur in response to respiratory acidosis (Section 7.7), but the concern here is not particularly with acid–base balance as such. The point is that the movement of the sodium, potassium and bicarbonate may be expected to lead to osmotic movement of water in the same direction. Whether the concentration of sodium in the extracellular fluid rises or falls depends on the relative amounts of sodium and water that enter it.

Assuming that osmotic equilibrium is restored after these movements of ions and water, we may think of the process as the movement of an isosmotic solution out of the cells. Let us take the osmolality of this solution, of the cells, and of the extracellular fluid as 300 mosmol/kg water.

Suppose now that 3 mmol of sodium moves from the cells into each kilogramme of extracellular water. Given that these sodium ions are accompanied by 5 mmol of other ions (potassium plus bicarbonate), a total of 8 mosmol of assorted ions enters each kilogramme of extracellular water. (Here, with trivial error, we equate moles and osmoles.) We are assuming that these solutes, accompanied by water, move effectively as a solution containing 300 mosmol/kg water and this means that the 8 mosmol take with them 27 g of water (i.e. $8 \times 1000/300$). This, then, is the amount of water accompanying the 3 mmol of sodium into each kilogramme of extracellular water.

6.3.1 **If 3 mmol of sodium moves from the cells into each kilogramme of extracellular water accompanied by 27 g of water, and if the initial extracellular sodium concentration is 150.0 mmol/kg water, what is the final concentration of sodium?**

It is evident that shifts of sodium between extracellular and intracellular fluid cannot be assessed on the basis of changing extracellular concentrations; it is also necessary to know what is happening to the extracellular fluid volume.

The total sodium content of the extracellular fluid is calculable as the product of its volume (or the mass of its water) and the sodium concentration; if both of these can be measured exactly, then the amount of sodium gained or lost can be calculated as a difference between two products. The extracellular fluid volume may be estimated as the volume of distribution of a substance, such as inulin, that spreads evenly throughout the volume and that does not enter the cells.

Difficulties remain. The 'inulin space' is subject to error and requires correction for the loss of inulin in the urine. The sodium concentration, best expressed in molal, rather than molar, terms (Section 6.8), is hard to measure to the accuracy called for in question 6.3.1. Moreover, the plasma and interstitial fluids should differ slightly in molal sodium concentration because of the Donnan effect (Section 6.6).

There is another general point to be made. Whenever sodium moves between cells and extracellular fluid with an equivalent amount of univalent anion (not necessarily accompanied by potassium), then water moves too and the extracellular concentration of sodium hardly alters. The concentration in the extracellular fluid cannot be significantly adjusted merely by shifts of sodium plus anion, but must be regulated by other means. It is a different matter when sodium crosses the cell membranes in 1:1 exchange for potassium, for then there is no water movement.

6.4 The role of bone mineral in the regulation of extracellular calcium and phosphate

The continual precipitation and dissolution of bone mineral is an important factor in the homeostasis of extracellular calcium and inorganic phosphate. It is modulated by such hormones as parathyroid hormone and calcitonin. Bone mineral consists mainly of four forms of calcium phosphate. These are their formulae:

$$Ca_9Mg(HPO_4)(PO_4)_6,$$

$$Ca_{8.5}Na_{1.5}(PO_4)_{4.5}(CO_3)_{2.5},$$

$$Ca_9(PO_4)_{4.5}(CO_3)_{1.5}(OH)_{1.5},$$

$$Ca_8(HPO_4)_2(PO_4)_4 \cdot 5H_2O.$$

The first three have very low solubilities and are far from being in equilibrium with extracellular fluid. The fourth is octocalcium phosphate. This is much the most soluble and is in equilibrium with bone extracellular fluid and close to equilibrium with the rest of the extracellular fluid.

The relationship amongst the solubility product for octocalcium phosphate and the concentrations of the relevant ions is one that obviously can be approached quantitatively, though not very simply, but that is not the intention here. Suffice to say that the solubility product for octocalcium phosphate has evidently been one of the determinants of extracellular fluid composition during evolution of the mammals.

It is a different feature of this composition that we now consider. This is the similarity between the concentrations of inorganic phosphate ($HPO_4 + H_2PO_4$) and of free calcium ions. Both are close to 1.3 mmol/l, so that the ratio Ca/P is close to unity. This fact is not predictable from the solubility relationship, for either concentration could be much higher, provided that the other were correspondingly lower. That ratios very different from unity are possible in extracellular fluids is illustrated by gastropods and other invertebrate animals. In many of these, the calcium concentration exceeds 10 mmol/l, while the concentration of inorganic phosphate is only 0.1–0.3 mmol/l. Invertebrates do not have bone mineral, but may contain other forms of precipitated calcium phosphate.

6.4.1 **For the above four bone salts, what is the range of values of the mole ratio Ca/P?**

The point to be made next depends on the similarity of these ratios to the Ca/P ratio in mammalian extracellular fluid. They are not identical, but they are clearly much more similar to each other than they are to the ratios of 30–100 in the above invertebrates. The following three calculations are about the effects on extracellular fluid of the dissolution and precipitation of bone mineral. Assume for this calculation that there are no exchanges with other body compartments and no change in the extent of binding of calcium to proteins.

6.4.2 **Suppose that calcium and phosphate dissolve into mammalian extracellular fluid from bone mineral in a ratio of, say, 1.5 Ca to 1 P – to the extent that the concentration of inorganic phosphate rises from 1.0 mmol/l to 1.3 mmol/l (i.e. by 30%). What is the percentage rise in free calcium, if the initial value is 1.3 mmol/l?**

To appreciate the point of this, consider now a hypothetical bony animal with concentrations as follows: free calcium, 10 mmol/l; inorganic phosphate, 0.1 mmol/l (as in some of the invertebrates). Assume, in addition, that calcium and phosphate dissolve from the mineral, or precipitate onto it, in the ratio of 1.5 Ca to 1 P and that there are no exchanges with other compartments of the body.

6.4.3 **If the concentration of inorganic phosphate is again raised by 30%, by what percentage is the concentration of calcium raised this time?**

6.4.4 **If, improbably, all the phosphate were to precipitate, what would be the final concentration of calcium in the extracellular fluid of that same hypothetical animal?**

From a comparison of these answers, it is evident that bone mineral can 'buffer' both ions effectively only in the mammal. In the hypothetical animal, the mineral could only contribute much to the homeostasis of calcium if huge variations in dissolved phosphate were tolerated – or else prevented by additional mechanisms such as renal regulation. The similarity of calcium and phosphate concentrations in mammalian extracellular fluid can thus be seen as advantageous. Whether this helped to mould the evolution of human body fluids is an open question, however. It must also be said that the similarity of calcium and phosphate concentrations also gave rise to a problem in homeostasis. Because calcium and phosphate necessarily move together between extracellular fluid and bone mineral, the other mechanisms that regulate them must be integrated accordingly. Note, for example, that when parathyroid hormone acts to raise extracellular calcium, it both mobilizes phosphate from bone mineral and increases the rate at which phosphate is excreted by the kidneys.

6.5 The principle of electroneutrality

According to this principle, in any solution the total charge on all the anions present is virtually equal to the total charge on all the cations. In other words, the concentration of anions in terms of equivalents is virtually equal to the concentration of cations in the same units. (For each ion in a solution, the number of equivalents present is equal to the number of moles multiplied by the respective valency.) This is a key idea in electrolyte and acid–base physiology, but it is given curiously little emphasis in many elementary accounts and seems sometimes to be forgotten even by research workers. Examples of its application are given below, but first that word 'virtually' must be considered.

The existence of a membrane potential implies that there is a slight imbalance of anions and cations across the cell membrane. When, as is typical, the interior of a cell is negatively charged with respect to the exterior, then there is a slight excess of anions over cations inside. Just how much of a discrepancy there is depends on the magnitude of the membrane potential (V_m, in volts) and membrane capacitance (C, in farad/cm^2). In terms of mol/cm^2 of membrane, the discrepancy is given by CV_m/zF. z is the valency of the ions, which will be considered here as all univalent (so that $z = 1$ and 1 mol = 1 equivalent). F is the Faraday constant. This is the specific ionic charge and is 96 490 coulomb/equiv, but here the round-number value of 10^5 coulomb/equiv (or 10^5 coulomb/mol, since $z = 1$) suffices.

6.5.1 **For a representative cell – with membrane capacitance, C, of 10^{-6} Farad/cm^2 (1 μF/cm^2) and a (negative) membrane potential of 0.07 volt (70 mV) – what is the discrepancy between the concentrations of anions and cations (in mol/cm^2), calculated as CV_m/zF?**
(1 farad = 1 coulomb/volt)

The answers needs now to be expressed in terms of mmol/l, but it then depends on the size and shape of the cell. Consider a length of cylindrical nerve fibre of radius r and length L. The surface area and volume are respectively $2\pi rL$ and $\pi r^2 L$. Combined with the last answer, these yield the following equation, where r is in μm:

$$\text{Discrepancy in mmol/l} = 14 \times 10^{-6}/r. \qquad (6.1)$$

6.5.2 **What is this discrepancy between anions and cations in (a) a nerve fibre of radius 10 μm, and (b) a nerve fibre of radius 0.5 μm?**

Compare these minute quantities with the typical intracellular potassium concentration of about 150 mmol/kg water. Would it be possible to demonstrate the discrepancy by chemical analysis? There are many other kinds of ion present and it is virtually impossible to draw up a detailed and accurate balance sheet for all the anions and cations in a particular cell or population of cells. Among the difficulties involved are those of identifying all the organic substances present and, when analysing a piece of tissue, of making exact allowance for the presence of extracellular fluid.

The movements of charge associated with action potentials are likewise tiny. Indeed their minuteness is important in allowing rapid changes in the value of V_m, and then the re-establishment of ionic gradients with the minimum expenditure of energy.

A sample of human plasma is analysed and found to contain ions at the following concentrations (mmol/l):

Na	144;	Cl	102;
K	4;	HCO_3	28;
free Ca	1;	lactate	1;
free Mg	0.5.		

The concentration of protein (which has a net negative charge) is estimated as 18 mequiv/l. (Since the number of equivalents of each ion is equal to the number of moles multiplied by the respective valency, 18 mequiv of protein carry the same charge as 18 mmol of univalent anion.)

6.5.3 **In units of mequiv/l, what is the difference between the total measured concentration of cations in this plasma sample and the total estimated concentration of anions?**

That the difference is not zero may be partly due to analytical error, but a small discrepancy is to be expected anyway, since some of the minor ionic constituents, such as sulphate and inorganic phosphate, have not been measured.

Another sample of blood plasma is analysed in another laboratory. This time it is from a Transylvanian blue-eyed bloater, a fish that has never previously been studied. The analysis, less complete, is as follows (with concentrations in mmol/l):

Na	138;	total Mg	3;
K	3;	Cl	70;
total Ca	1.5;	HCO_3	5.

6.5.4 **Again in mequiv/l, what is the discrepancy between measured cations and anions? What could it mean?**

The example is imaginary, but it is not unlike many analyses that have been published without relevant comment. The discrepancy is large, but its meaning is not clear. Does the plasma contain a high concentration of (negatively charged) protein or other anion? Are the analyses faulty? What is clear is that there is something to be investigated further. It often pays to check the balance of anions and cations,

Human erythrocytes (in arterial blood) contain sodium, potassium, chloride and bicarbonate at approximately the following concentrations (mmol/kg water):

Na	18;	Cl	78;
K	135;	HCO_3	16.

Free calcium is negligible in the context of charge balance. The concentration of free magnesium is about 0.5 mmol/kg water.

6.5.5 **What is the discrepancy between the concentrations of anions and cations, in mmol/kg water? How might it be interpreted?**

6.6 Donnan equilibrium

The glomerular filtrate has been described as resembling plasma that is nearly free of its proteins (except that any substances bound to the proteins would stay with them in the plasma). More detailed consideration reveals that there must be other minor differences in ionic composition between plasma and filtrate, these being due to the net negative charge on the plasma proteins. This brings us to the topic of the Donnan (or Gibbs–Donnan) equilibrium. This is also relevant to the filtration of fluid through the walls of capillaries generally, and to the experimental dialysis of protein solutions. Although the subject seems sometimes to be over-emphasized in elementary teaching, it can be worth looking at for the use made of the Nernst equation and the principle of electroneutrality. It also has an important bearing on colloid osmotic pressures (Section 6.7).

Consider any two solutions that are separated by a membrane that is permeable to all ions present except proteins. It could, for example, be a cellulose dialysis membrane or, less ideally, capillary endothelium. The permeant ions are allowed to reach diffusion equilibrium. (As is explained below, there is likely to be a difference in osmotic pressure between the two solutions, this being the colloid osmotic pressure. To prevent water movement, there would therefore need to be a small opposing difference in hydrostatic pressure.)

A simple rule relates the various concentrations of diffusible ions:

$$\frac{[\mathrm{Na}]_1}{[\mathrm{Na}]_2} = \frac{[\mathrm{Cl}]_2}{[\mathrm{Cl}]_1} = \frac{\sqrt{[\mathrm{Ca}]_1}}{\sqrt{[\mathrm{Ca}]_2}} \cdots \text{ and so on.} \tag{6.2}$$

The subscripts 1 and 2 refer to the two solutions.

If this set of relationships is unfamiliar, note that it may be derived by equating the equilibrium potentials for all ionic species present, except protein, these potentials being given in each case by the Nernst equation. Thus, for sodium and chloride at 37 °C,

$$E_{\mathrm{Na}} = 61.5 \log \frac{[\mathrm{Na}]_1}{[\mathrm{Na}]_2} = E_{\mathrm{Cl}} = 61.5 \log \frac{[\mathrm{Cl}]_2}{[\mathrm{Cl}]_1}. \tag{6.3}$$

Now divide through by 61.5 and take antilogarithms and you have the first part of equation (6.2).

For simplicity, let us take solution (1) as containing only NaCl, at concentration C mmol/kg water. Solution (2) contains NaCl too, but it also contains protein at concentration E mequiv/kg water. In accordance with the principle of electroneutrality,

$$[\mathrm{Na}]_1 = [\mathrm{Cl}]_1 = C \tag{6.4}$$

and

$$[\mathrm{Na}]_2 = [\mathrm{Cl}]_2 + E. \tag{6.5}$$

From equations (6.2), (6.4) and (6.5),

$$([\mathrm{Cl}]_2 + E)[\mathrm{Cl}]_2 = C^2. \tag{6.6}$$

Thus (an equation to apply here, but not remember),

$$2[\mathrm{Cl}]_2 = \sqrt{(E^2 + 4C^2)} - E. \tag{6.7}$$

Consider now a pair of solutions such that C is 150 mmol/kg water and E is 18 mequiv/kg water, values chosen to be like those in human plasma. $[\mathrm{Cl}]_2$, the concentration of chloride in the solution with the protein, works out as $[\sqrt{\{18^2 + (4 \times 150^2)\}} - 18]/2 = 141.3$ mmol/kg water.

6.6.1 **According to equation (6.5), what is $[\mathrm{Na}]_2$?**

6.6.2 **As a check, is $[\mathrm{Na}]_1/[\mathrm{Na}]_2$ equal to $[\mathrm{Cl}]_2/[\mathrm{Cl}]_1$, in accordance with equation (6.2)?**

The general magnitude of this 'Donnan ratio' for plasma may be worth remembering – close to unity, yet readily distinguishable from it by chemical analysis. Both ($[Na]_2 - [Na]_1$) and ($[Cl]_1 - [Cl]_2$) work out nearly equal to $E/2$ and this is a convenient conclusion too.

In practice, it is not usually fruitful to apply these equations to plasma with the expectation of a precise answer. The value of E is rarely known exactly and even equation (6.2) tends not to fit analytical data, partly because of the specific binding of ions to proteins. (What is more, the difference between binding and Donnan effects can be difficult to separate, even conceptually.)

One other aspect of the Donnan equilibrium is, however, worth quantifying before we go on to look at the relevance of the equilibrium to colloid osmotic pressure. This is the electrical potential between the two solutions. Since both sodium and chloride ions are at equilibrium, the electrical potential is equal to both E_{Na} and E_{Cl} in equation (6.3), this being the Nernst equation formulated in terms of mV at 37 °C. $[Na]_1/[Na]_2$ and $[Cl]_2/[Cl]_1$ have already been calculated.

6.6.3 **Solution 2, which contains the (negatively charged) protein, is slightly negative with respect to solution 1. What is the electrical potential difference?** ($\log 0.942 = -0.026$)

6.7 Colloid osmotic pressure

The colloid osmotic pressure, or oncotic pressure, of a solution is that part of the total osmotic pressure that is due to colloids. In natural body fluids the colloids are the proteins. As a round number, the colloid osmotic pressure of human plasma is about 25 mmHg.

It is important to realize how very different this value is from the total osmotic pressure but, because the two are usually expressed in different units, the point is often unappreciated. The total osmotic concentrations of the body fluids are not usually given in pressure units, but as osmolalities (see Notes and Answers). As a round number, the osmolality of most human body fluids (cells and extracellular fluid) is 300 mosmol/kg water. As discussed in Notes and Answers, 1 mosmol/kg water exerts an osmotic pressure of 19.3 mmHg at body temperature.

6.7.1 **Given that relationship, and assuming an osmolality of 300 mosmol/kg water, what is the osmotic pressure of human body fluids in units of mmHg?**

Compare this with the colloid osmotic pressure of human plasma of about 25 mmHg.

It might be supposed that colloid osmotic pressures would be related to protein concentrations in accordance with the relationship given above for other solutes. In general, the colloid osmotic pressure does increase with the concentration of protein and, for a given concentration in g/l or g/kg water, proteins of low molecular weight (e.g. albumins) do exert a higher osmotic pressure than those of high molecular weight (e.g. globulins). In short, it is roughly true to say that the colloid osmotic pressure increases with the concentration of protein in terms of mmol/kg water. This may be as much as most physiologists need to know on the subject, but it is an easy matter to probe a little further making use of data obtained already in Section 6.6.

Fig. 6.1 summarizes the data of question 6.6.2. Solutions 1 and 2 are separated by a membrane that is permeable to everything except protein. The solutions are in Donnan equilibrium. Solution 1 contains only NaCl and solution 2 contains protein at a concentration similar to that of blood plasma.

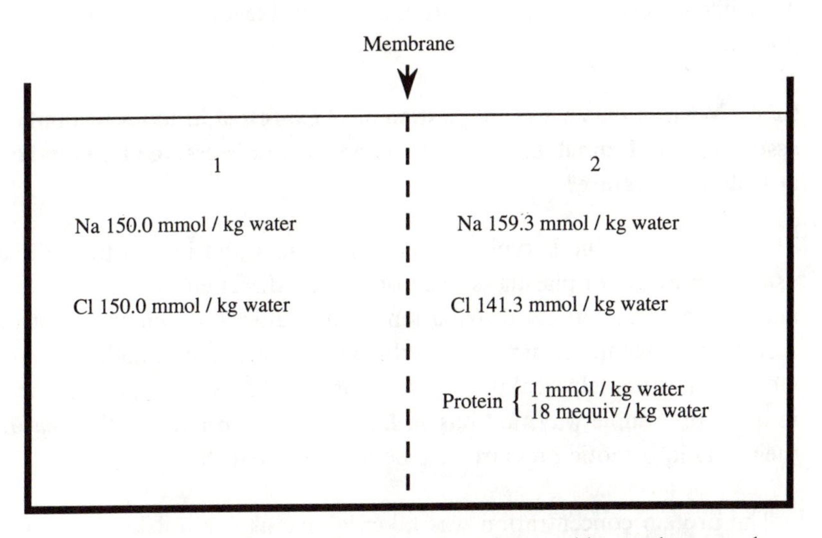

Fig. 6.1. Solutions 1 and 2 are separated by a semipermeable membrane and are in Donnan equilibrium. Sodium and chloride are present at the concentrations shown. Proteins are present in solution 2 only and give rise to a colloid osmotic pressure that tends to draw water from left to right.

6.7.2 **In which solution is the sum of concentrations ([Na] + [Cl]) greater, the one with the protein (2) or the one without it (1)? By how much is the sum greater?**

It is evident that the protein in solution 2 raises the osmotic pressure above that in solution 1 both directly and indirectly. The 'colloid osmotic pressure' of solution 2 is not only due to the direct effect of the proteins themselves, but also to the slight excess of ions that they attract.

To quantify these effects it suffices here to regard 1 mosmol of protein, sodium or chloride as the same as 1 mmol, as in an 'ideal' solution. Accordingly, 1 mmol/kg water of any of these exerts an osmotic pressure of 19 mmHg at body temperature.

In the calculation of the sodium and chloride concentrations in solution 2, the concentration of protein had to be expressed in terms of charge, i.e. as 18 mequiv/kg water. It must now be specified in terms of mmol and a reasonable value to take is 1 mmol/kg water.

6.7.3 **In mmol/kg water, how much more solute (i.e. protein + sodium + chloride) is there in solution 2 as compared with solution 1?**

This difference in total concentration is responsible for the colloid osmotic pressure of solution 2.

6.7.4 **What is the answer to question 6.7.3 expressed in terms of mmHg, assuming that 1 mmol/kg water exerts an osmotic pressure of 19 mmHg at body temperature?**

The calculated value is typical of human plasma and indeed the colloid osmotic pressure of plasma is due both to the direct effect of the protein and to the asymmetrical distribution of inorganic ions, whether that be across endothelium or across the artificial membrane of a colloid osmometer. Unfortunately, real protein solutions do not behave quantitatively in quite the simple way that has so far been implied and for this reason their colloid osmotic pressures are better measured than calculated.

The protein concentration was taken above as 1 mmol/kg water, but more usual units for protein are g/l or g/kg water. For any given protein, the concentration in g/kg water is calculable as the concentration in mmol/kg water multiplied by one-thousandth of the molecular weight.

6.7.5 **A solution contains protein at 1 mmol/kg water, as above. What would the concentration be in g/kg water if the protein were albumin of molecular weight 68 000?**

The protein content of plasma is typically 65–80 g/kg water, with about 60% being albumin. About 35% is globulin. The globulin has a much higher average molecular weight and so contributes proportionately less to colloid osmotic pressure.

6.8 Molar and molal concentrations

Concentrations of solutes in body fluids may be expressed in terms of millimolarity (mmol/l of solution, mM) or millimolality (mmol/kg of water). In extracellular fluid the numerical difference is small enough that for many of the calculations in this book it is unimportant. Moreover, in an elementary course of physiology the difference may well be regarded as a distracting complication that is best ignored. Nevertheless, the matter is important to the interpretation of actual clinical measurements on plasma and with cytoplasm the difference between the two measures is large.

Let us start by defining a conversion factor, W, such that

$$\begin{aligned} W &= \text{millimolar concentration/millimolal concentration} \\ &= \frac{\text{mmol/l solution}}{\text{mmol/kg water}} \\ &= \text{kg water/l solution, or g water/ml solution.} \end{aligned} \qquad (6.8)$$

Consider now a salt solution that contains proteins at concentration c g/l. Let W have a value of W_0 when c is zero. In the case of human plasma without its proteins, W_0 is about 0.99 kg water/l solution. Let the volume occupied by a gram of protein be V ml (V being the 'partial specific volume' of the protein). For most proteins, V is 0.70 to 0.75 ml/g; for plasma proteins it is about 0.75 ml/g. The various quantities are related thus:

$$W = W_0(1 - Vc/1000), \qquad (6.9)$$

or, for blood plasma,

$$W = 0.99(1 - 0.75c/1000). \qquad (6.10)$$

The equation is not worth remembering, but the calculating physiologist may find it useful that it is given here.

6.8.1 **What is W for plasma with a (typical) protein concentration of 70 g/l?**

6.8.2 **If plasma with that protein content contains sodium at 141 mmol/l, what is the concentration of sodium in mmol/kg water?**

Since 141 mmol/l is within the normal range for sodium of about 137–145 mmol/l, the calculated concentration in mmol/kg water can be taken as a convenient round-number value for plasma in general.

So far, one component of plasma has been ignored that can be very relevant; this is the lipids. Normal concentrations in adults are about 5–9 g/l. Lipids make no difference to the millimolal concentrations of other substances, but lower the millimolar concentrations of these substances by adding to the total volume. In cases of hyperlipaemia, there may therefore be an apparent hyponatraemia when the sodium concentration is recorded in units of mmol/l – even though the concentration in mmol/kg water is normal. It is the concentration in units of mmol/kg water that matters so far as cell function is concerned.

In practice, one might wish to effect the conversion from millimolar to millimolal on the basis of two measurements that were not used above, namely the density of the solution and its water content (g water/g solution) as determined by weighing and drying. Then,

$$W = \text{density (g/ml)} \times \text{water content (g/g)}. \qquad (6.11)$$

In cytoplasm, the difference between millimolar and millimolal concentrations is much greater, because of the large amount of substance present that is not water. Those who analyse tissues generally do so on a basis of mass rather than of volume and end up neither with concentrations in units of mmol/l of cytoplasm, nor with the information necessary for the calculation of these. Some authors of textbooks, on the other hand, do write, loosely, of mM (which, strictly, is an abbreviation for mmol/l). Since millimolar concentrations are rarely meaningful in this context, we may, to illustrate the difference between units, be content with just a rough calculation.

For this purpose we need a representative value of c, but this time including all the organic components of the cytoplasm and not only the protein. For erythrocytes, a figure of 340 g/l may be familiar as the 'mean corpuscular haemoglobin concentration' (MCHC); c in erythrocytes is only slightly higher – about 360 g/l. This is also about right for some other cells.

6.8.3 **Treating cytoplasm as if it were like very concentrated blood plasma, and applying equation (6.10), what is the conversion factor, W, if c is 360 g/l?**

6.8.4 **If a cell contains potassium at a concentration of 150 mmol/kg water, what is the concentration in mmol/l, assuming the value of W just calculated?**

6.9 Osmolarity and osmolality

Whereas the mole and millimole are usual units for individual solutes, the osmole and milliosmole, defined in Notes and Answers, are generally used only in the context of total solute concentration, and especially where that relates to osmotic pressure. The difference between osmolarity and osmolality is the same as the difference between molarity and molality; osmolarity is the concentration in osmol/l of fluid and osmolality is the concentration in osmol/kg water. As with molarity and molality, the distinction, simple though it is, can be more troublesome than useful in an elementary course in physiology.

Whilst osmolality is directly measurable in terms of freezing point or vapour pressure, osmolarity cannot be measured directly and is not generally a useful concept. If people tend to speak more of osmolarities, perhaps it is because the word is more euphonious.

One context in which it does arguably make sense to think in terms of osmolarities rather than osmolalities is in comparing the osmotic concentration of a solution with the millimolar concentrations of the various contributory solutes. One might, for example, add up all the solutes given for the plasma sample of question 6.5.3 to see if they are roughly compatible with normal values of osmolality/osmolarity. Here, again, are those concentrations (in mmol/l):

Na	144;	Cl	102;
K	4;	HCO_3	28;
free Ca	1;	lactate	1;
free Mg	0.5.		

The protein concentration can be taken as about 1 mmol/l.

6.9.1 **What is the sum of all these concentrations?**

A round-number value for osmolality of 300 mosmol/kg water was used for plasma in the calculations above. Actual values are usually in the range of 290–300 mosmol/kg. water. Considering that the units are different, the match could be regarded as reasonable, especially since other solutes, such as glucose and urea, are not included in the total obtained in question 6.9.1. For a more accurate comparison, corrections are needed for (1) the difference between osmolarity and osmolality, and (2) the non-ideal behaviour of solutions, i.e. the fact that 1 mmol of solute is generally less than 1 mosmol.

At this point one may be tempted to skip the tedium of making such corrections for what is only a hypothetical example – and to skip, therefore, the next calculation. The reward for reading on is the demonstration that calculation happens often to be unnecessary.

To convert the above answer to mmol/kg water without knowing the exact protein concentration in g/l, one may choose to apply the value of W calculated in the answer to 6.8.1, dividing therefore by 0.94. To convert mmol to mosmol, the answer needs then to be multiplied by an empirical factor called an 'osmotic coefficient'. This depends on the nature and concentrations of the solutes, but for a pure solution of NaCl, 150 mmol/kg water, the coefficient is 0.93. This value is about right for plasma.

6.9.2 **What is the combined correction factor, to one decimal place?**

Thus, for typical blood plasma, the concentration in mosmol/kg water happens to be nearly equal to the sum of all the solute concentrations in mmol/l.

6.10 Gradients of sodium across cell membranes

Sodium is at a much lower concentration inside a typical cell than it is outside. Because of this gradient and the negative membrane potential, sodium tends to leak in, but its concentration is kept low inside by the continual bailing action of the sodium pump (Na, K-ATPase). Typically, three sodium ions are transported outwards for each molecule of ATP that is hydrolyzed, and two potassium ions are transported into the cell.

There is an upper limit to the electrochemical potential gradient for sodium that the pump can maintain and this is determined by the amount of energy that is available from each mole of ATP. Potassium is also relevant, since this ion is transported by the sodium pump, too. However, the argument and calculations are easier if we disregard this ion (which we do only for the moment), and the conclusions are not greatly affected.

The electrochemical potential difference for sodium consists of two terms. One is the membrane potential (V_m, negative inside) and the other is the equilibrium potential for sodium, E_{Na}. Thus,

$$\text{electrochemical potential difference} = E_{Na} - V_m. \tag{6.12}$$

E_{Na} is related by the Nernst equation to the extracellular and intracellular concentrations (per kg water) of sodium, these being $[Na]_e$ and $[Na]_i$ respectively:

$$E_{Na} = RT/F \ln([Na]_e/[Na]_i). \tag{6.13}$$

R is the gas constant, F is the Faraday constant and T is the absolute temperature. For a body temperature of 37 °C the equation may be rewritten more conveniently as

$$E_{Na} = 61.5 \log([Na]_e/[Na]_i). \tag{6.14}$$

This has appeared before, as part of equation (6.3). As a realistic example, suppose that $[Na]_e$ is 150 mmol/kg water and $[Na]_i$ is 15 mmol/kg water. Then, E_{Na} is 61.5 log 10 = 61.5 mV.

To return to the electrochemical potential difference, this is also related to the thermodynamic work, W, needed to transport 1 mol of sodium across the membrane, as follows

$$E_{Na} - V_m = W/F. \tag{6.15}$$

If it is assumed now that the sodium pump acts to maintain the maximum electrochemical gradient, then we can relate the latter to the amount of energy available from the hydrolysis of ATP. Assuming that one ATP molecule powers the transport of three sodium ions, W is equal to one-third of $-\Delta G_{ATP}$, the free energy change in the hydrolysis of ATP. Hence, equation (6.15) may be rewritten as

$$E_{Na} - V_m = -\Delta G_{ATP}/3F. \tag{6.16}$$

The value of $-\Delta G_{ATP}$ depends on the concentrations of the various reactants in the cytoplasm and it varies from cell to cell. It may, however, be taken as about 10–13 kcal/mol. The maximum electrochemical potential difference, ($E_{Na} - V_m$), thus works out, in volts, at 10–13 kcal/mol divided by $3F$. The value of F is 23.1 kcal/volt·equiv. In the case of sodium, 1 equiv = 1 mol.

6.10.1 **What is the likely range, in millivolts, for the maximum electrochemical potential difference?**

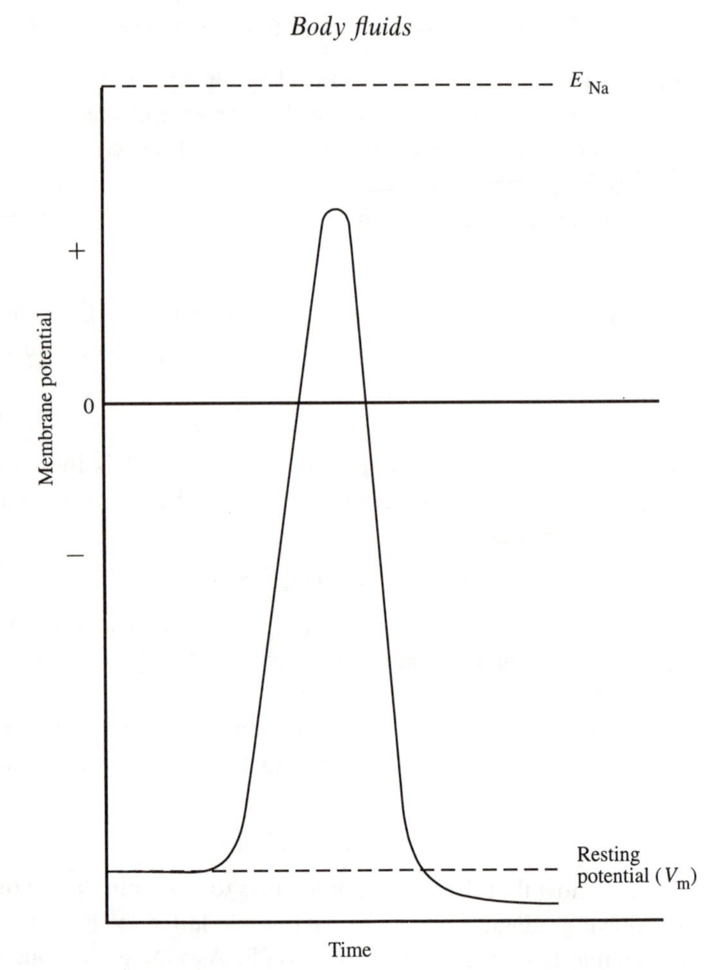

Fig. 6.2. Diagram of a sodium action potential in a nerve fibre. V_m is the resting membrane potential and E_{Na} is the equilibrium potential for sodium.

Let us now see how close the actual electrochemical potentials for sodium are to this range, calculating them from ($E_{Na} - V_m$). For a quick and approximate answer, achieved without having to think about logarithms, it may be recalled that the peak positive potential in a sodium action potential (the overshoot) approaches (without actually reaching) the equilibrium potential for sodium (E_{Na}). This is shown in Fig. 6.2.

6.10.2 **A textbook gives −90 mV as the resting potential for a nerve fibre and +40 mV for the overshoot of the action potential. What estimate does this yield for the electrochemical potential difference for sodium?**

This is obviously a rather cavalier approach to the problem, but the answer is not far from the 144–188 mV calculated above for the maximum that might be produced by the action of the sodium pump. Remember that the peak of the action potential does not actually achieve the sodium equilibrium potential.

Now let us approach the matter in the more obvious way, calculating the value of E_{Na} from concentrations of sodium (equation 6.14). To keep the arithmetic easy, let us start by postulating external and internal concentrations of, respectively, 150 and 15 mmol/kg water. Let the resting membrane potential be -90 mV again.

6.10.3 **What is now the estimate for ($E_{Na} - V_m$)?**

The answer is within the previously calculated range of 144–188 mV. Note, however, that many estimates for intracellular sodium concentration are lower than our convenient choice of 15 mmol/kg water. Note also that chemical estimates of total sodium (e.g. by flame photometry) may include some that is not actually free in the cytoplasm, whereas $[Na]_i$ in equation (6.13 and 6.14) refers to free sodium.

6.10.4 **What would be the new estimate of ($E_{Na} - V_m$), if the internal concentration were halved?** (For log 2, see Appendix A)

Is this new estimate also within the range of 144–188 mV calculated above as the maximum achievable by the sodium pump?

Nobody would take these calculations as implying that the electrochemical potential difference for sodium across all cell membranes is exactly that which can be sustained by sodium pumps transporting three sodium ions per molecule of ATP hydrolysed, but the correspondence is close. In any case, the continual influx of sodium ions, accelerated during action potentials, would tend to keep the electrochemical difference below the theoretical maximum.

The rest of this section is for those who prefer rigour to simplicity; there are no surprises. So far we have noted only in passing the fact that the sodium pump exchanges sodium for potassium. Equation (6.16) may now be revised accordingly. It is assumed that for every three sodium ions transported there are two potassium ions transported, and therefore one net negative charge. The equilibrium potential difference for potassium, E_K, is now required in the calculation. As with sodium, this is given by the Nernst equation, which, for 37 °C, is:

$$E_K = 61.5 \log([K]_e/[K]_i). \qquad (6.17)$$

Thus, for example, for $[K]_e = 4.5$ and $[K]_i = 150$, E_K is -94 mV.

The new version of equation (6.16) is as follows:

$$-\Delta G_{ATP}/3F = E_{Na} - \tfrac{2}{3}E_K - \tfrac{1}{3}V_m. \qquad (6.18)$$

6.10.5 **What is the value of the right-hand side of equation (6.18) if V_m is -90 mV and the ionic concentrations, in mmol/kg water, are as before: $[Na]_e = 150$; $[Na]_i = 15$; $[K]_e = 4.5$; $[K]_i = 150$?**

Taking the transport of potassium into account thus leads to much the same quantitative conclusion as before (see especially the answer to 6.10.3, where the internal sodium concentration is the same). This is because the value of V_m is often close to that of E_K. However, E_K is, in fact, typically more negative than V_m, as above, so that $(E_{Na} - \tfrac{2}{3}E_K - \tfrac{1}{3}V_m)$ is somewhat greater than $(E_{Na} - V_m)$.

6.11 Membrane potentials — simplifying the Goldman equation

Teachers and textbook writers disagree on whether the Goldman equation (or 'constant field equation') is necessary for a basic understanding of membrane potentials. The main aim here is familiarization and simplification. Here is one form of the equation, with V_m being the membrane potential (in millivolts) and the value 61.5 being a constant appropriate at 37 °C (as in equation 6.14):

$$V_m = 61.5 \log \frac{[K]_e + \alpha[Na]_e}{[K]_i + \alpha[Na]_i}. \qquad (6.19)$$

The equation may be derived in more than one way, with the meaning of α varying accordingly. This is explained in the next two paragraphs, with minimum mathematical detail.

The Goldman equation is usually given in a form much like this:

$$V_m = \frac{RT}{F} \cdot \ln \frac{P_K[K]_e + P_{Na}[Na]_e + P_{Cl}[Cl]_i}{P_K[K]_i + P_{Na}[Na]_i + P_{Cl}[Cl]_e}$$

$$= 61.5 \log \frac{P_K[K]_e + P_{Na}[Na]_e + P_{Cl}[Cl]_i}{P_K[K]_i + P_{Na}[Na]_i + P_{Cl}[Cl]_e}. \qquad (6.20)$$

P_K, P_{Na} and P_{Cl} are membrane permeability coefficients for potassium, sodium and chloride. The equation is valid when there is no net current flowing through the membrane. (Additional terms can be included to allow for the diffusion of other ions through the membrane, but these are not usually significant.) Making the equation more like that in the first paragraph (equation 6.19), the terms for chloride are commonly omitted, and this is appropriate when these ions are close to equilibrium across the membrane. This they often are, and then they have no tendency to carry net current inwards or outwards. Omitting the chloride terms and dividing top and bottom by P_K, we have:

$$V_m = 61.5 \log \frac{[K]_e + P_{Na}/P_k[Na]_e}{[K]_i + P_{Na}/P_K[Na]_i}. \tag{6.21}$$

With the term α standing for the permeability ratio P_{Na}/P_K, this is equivalent to equation (6.19) above.

Equation (6.21) takes no account of the net ionic current through the cell membrane due to the sodium–potassium pump. There is a net ionic current because only two potassium ions are carried inwards for every three sodium ions transported outwards. There is, however, an alternative derivation of equation (6.19) that does involve this current, although α then has a different meaning. The derivation is mathematically similar to that of the usual Goldman equation and is not detailed here. It starts from assumption that the net passive diffusional fluxes of potassium and sodium are each equal and opposite to their net non-diffusional fluxes. A steady state is therefore assumed. The ratio of net potassium efflux to net sodium influx, each by passive diffusion, is determined by V_m, P_{Na}/P_K and the ionic concentrations. If this ratio is f, then equation (6.19) applies, with α equal to fP_{Na}/P_K. The ratio f applies also to the non-diffusional fluxes and would be $\frac{2}{3}$ if these were entirely due to Na, K-ATPase. As to other mechanisms, Na–H countertransport and inward Na–K–2Cl cotransport would both raise the value of α. (Alternatively, these two mechanisms may be accommodated in separate terms.)

There are two important points to note about equation (6.19). First, when α is very low (i.e. the membrane is much more permeable to potassium than to sodium), equation (6.19) approximates to the Nernst equation for potassium (equation 6.17) and V_m therefore approaches the equilibrium potential for potassium, E_K. Second, when α is very high (i.e. the membrane is much more permeable to sodium than to potassium, as during the upstroke of the action potential), the equation approximates to the Nernst equation for sodium (equation 6.14) and the value of V_m approaches that of E_{Na} (as assumed in Section 6.10).

Common textbook values for α, (i.e. P_{Na}/P_K), for mammalian nerve and muscle, are 0.01 or 0.05. For some non-excitable cells the ratio is much higher, even 0.3–0.5 (with V_m sometimes less than -20 mV). Just so long as one does not deal with particular cells and circumstances, it matters little, because of that variability, exactly what α stands for.

For the sake of easy arithmetic, let us now seek further simplification.

6.11.1 **Suppose that $[K]_i = 150$ mmol/kg water and $[Na]_i = 15$ mmol/kg water. In relation to the bottom line in equation (6.19), what is $\alpha[Na]_i$ as a percentage of $[K]_i$ when α is (a) 0.05 and (b) 0.5?**

It is thus evident that equation (6.19) may be approximated for ordinary resting conditions as

$$V_m = 61.5 \log \frac{[K]_e + \alpha[Na]_e}{[K]_i}$$

$$= -61.5 \log \frac{[K]_i}{[K]_e + \alpha[Na]_e}. \qquad (6.22)$$

This is only valid when the values of α and $[Na]_i$ are not too high but, because of the logarithmic relationship, the effect on V_m of omitting the term $\alpha[Na]_i$ is further reduced.

For practice calculation, consider a cell for which $[K]_e$, $[Na]_e$ and $[K]_i$ are respectively 4.5, 150 and 150 mmol/kg water. Since graphs showing the dependence of V_m on $[K]_e$ are commonplace, let us vary α instead of $[K]_e$. (Variations in the ratio P_{Na}/P_K, and hence α, may be produced locally by transmitters such as acetylcholine.)

6.11.2 **What would V_m be for this cell if α were (a) zero, (b) 0.01 and (c) 0.07?** (log 3 = 0.477, i.e. about 0.5. log 4 = 2 log 2)

7
Acid–base balance

The most important quantitative relationships in acid–base physiology are (1) the definition of pH, (2) the Henderson–Hasselbalch equation (with which may be included the relationship between tension and concentration of dissolved carbon dioxide), and (3) the Principle of Electroneutrality which has already been treated in Chapter 6, Body Fluids. It is assumed that these are all to some extent familiar, but the essentials of (1) and (2), and of the subject of buffering, are re-stated here in the knowledge that acid–base balance is a blind spot for many physiologists. As usual, the hope has been to put a new slant on the subject and at the same time provide opportunities for the exercise of familiar concepts and quantities.

Going beyond what is mainly physical chemistry (Sections 7.1, 7.2, and 7.4 to 7.6), there are calculations on aspects of cell pH and bicarbonate (Section 7.3), the roles in whole-body acid–base balance of cells and bone mineral (Sections 7.7 and 7.8), and the postprandial alkaline tide (Section 7.9).

7.1 pH and hydrogen ion activity

S. P. L. Sørensen introduced the pH notation in 1909, with pH (originally P_H) standing for $-\log[H]$, [H] being the concentration of hydrogen ions in mol/l. Thus, pH 3 corresponds to 10^{-3} mol/l (1 mmol/l). Most people find this definition adequate for their purposes and for the moment we stay with it.

It is of course more natural, even if the numbers are sometimes cumbersome, to think in terms of concentrations rather than of negative logarithms. Physiologists who think about [H] in preference to pH may know the answer to the next question already. It is posed partly to make use of log 2 ≃ 0.3 (Appendix A), but the answer is used later. [H] can be calculated as the antilogarithm of −pH, or as $10^{-\mathrm{pH}}$.

7.1.1 **What is the approximate concentration (mol/l or nmol/l) of hydrogen ions in plasma at pH 7.4?**
$(10^{-7.4} = 10^{-8} \times 10^{0.6} = 10^{-8} \times 10^{(2 \times 0.3)})$

Sørensen's first definition runs into difficulties with very small volumes of fluid. Recalling that 1 mol of a substance contains 6.0×10^{23} molecules or ions (Avogadro's number), let us take the case of a solution of pH 7, for which [H] would be 10^{-7} mol/l.

7.1.2 **If a fluid contains hydrogen ions at a concentration of 10^{-7} mol/l, how many should there be in 1 μm³?** (1 μm³ = 10^{-15} l)

This answer may be better appreciated in relation to the volumes of cell organelles such as mitochondria. Mitochondria are typically about 0.2–1 μm across and 2–8 μm in length.

7.1.3 **What is the volume of a roughly cylindrical mitochondrion of diameter 0.2 μm and length 4 μm?** (Take π as 3, as in the Old Testament)

7.1.4 **Combining the results from 7.1.2 and 7.1.3, approximately how many hydrogen ions would such a volume of fluid contain if the pH were 7?**

Our mitochondrion is small, but not the smallest. Moreover, there is within the whole mitochondrion a much smaller space, that between the inner and outer membranes. Even more significant, the inner, matrix space of the mitochondrion is actually much more alkaline than the cytosol (the pH of which may itself exceed 7.0) and would therefore contain only a small fraction of the number of hydrogen ions calculated above. This internal alkalinity is due to the continual translocation of hydrogen ions across the inner membrane that is required to produce the gradient necessary for ATP synthesis. Contrast the paucity of hydrogen ions in the mitochondrion with their frequent to-ing and fro-ing across the inner membrane!

As a general conclusion, there would seem to be well-defined spaces within cells that contain only fractions of hydrogen ions. One is therefore forced to think of pH as reflecting, not actual concentrations, but something more like averages in time. The point is that hydrogen ions do not individually exist free for long (they are in any case hydrated in aqueous solutions), but are continually appearing and disappearing as they dissociate from, and reassociate with, hydroxyl ions and buffer molecules.

We have seen one difficulty in defining pH in terms of hydrogen ion concentration. There is a second difficulty: whereas it is easy to prepare solutions of known concentration of, say, sodium, this is not generally true of hydrogen ions. It is only possible to have direct and exact knowledge of [H] in strong solutions of strong acid. It is therefore necessary to define the pH of a solution operationally by comparison with standard solutions of buffer. The defined pH value of a buffer standard is allotted with the intention that it should approximate to $-\log(\mathrm{H})$, where (H) corresponds to something like the activity of hydrogen ions in the solution. Thus it is that the pH of 0.1 mol/l HCl is not 1.0, but 1.09 (at 25 °C). As the activity of a single ionic species is not strictly definable in theory, the whole concept of pH can be a little troublesome. However, a proper definition of pH is simply this – 'the reading on a correctly calibrated pH meter'. When used below, the term 'activity' in relation to hydrogen ions and the abbreviation (H) both mean $10^{-\mathrm{pH}}$.

7.2 The CO_2–HCO_3 equilibrium: the Henderson–Hasselbalch equation

The Henderson–Hasselbalch equation is fundamental to the quantitative treatment of acid–base balance. Here it is in its most commonly used form:

$$\mathrm{pH} = \mathrm{pK}'_1 + \log \frac{[\mathrm{HCO_3}]}{S \cdot P_{\mathrm{CO_2}}}. \tag{7.1}$$

pK'_1 is the equilibrium constant for the following reaction:

$$\mathrm{CO_2} + \mathrm{H_2O} = \mathrm{HCO_3^-} + \mathrm{H^+}. \qquad [7.1]$$

The value of pK'_1 in human blood plasma at 37 °C is about 6.1. S is the solubility coefficient for carbon dioxide and its value under the same conditions is about 0.03 mmol/l per mmHg. As noted in Section 4.2, the product $S \cdot P_{\mathrm{CO_2}}$ gives the concentration of dissolved carbon dioxide in mmol/l. For a normal arterial $P_{\mathrm{CO_2}}$ of 40 mmHg, this product is 1.2 mmol/l, being about one-twentieth of the concentration of bicarbonate. It is roughly seven hundred times the concentration of carbonic acid, which is therefore a trivial contributor to 'total carbon dioxide'; books having

'$[H_2CO_3]$' in place of '$S \cdot P_{CO_2}$' in equation (7.1) are therefore seriously wrong.

7.2.1 **At this P_{CO_2} of 40 mmHg, what is the pH when the concentration of bicarbonate is (a) 12 mmol/l and (b) 24 mmol/l?** (log 2 = 0.3)

The higher of the two bicarbonate concentrations is in the normal range for arterial plasma (22–30 mmol/l) – and of course this must be so since both the calculated pH and the P_{CO_2} are normal. Both 24 mmol/l and 25 mmol/l are used below as representative, this being partly for arithmetical reasons that should become evident.

For the free use of the Henderson–Hasselbalch equation with minimum calculation, it is a great help to appreciate that it has the following differential forms:

$$\Delta \text{pH} = \Delta \log[\text{HCO}_3] - \Delta \log P_{CO_2}, \tag{7.2}$$

or, for any two conditions 1 and 2,

$$\begin{aligned} \text{pH}_1 - \text{pH}_2 &= \log[\text{HCO}_3]_1 - \log[\text{HCO}_3]_2 - \log(P_{CO_2})_1 + \log(P_{CO_2})_2 \\ &= \log\frac{[\text{HCO}_3]_1}{[\text{HCO}_3]_2} + \log\frac{(P_{CO_2})_2}{(P_{CO_2})_1}. \end{aligned} \tag{7.3}$$

Note that pK_1' and S have conveniently disappeared.

7.2.2 **By how much does the pH change if the bicarbonate concentration is halved?** (log 2 = 0.3)

7.2.3 **To one decimal place, by about how much does the pH change if the P_{CO_2} rises from 39 to 83 mmHg?** (Again use log 2, for only an approximation is required)

Log 2 has been used four times so far in this chapter. The value of remembering it is stressed in Appendix A and in acid–base balance we have a context in which it can be especially useful. As just illustrated, one may immediately calculate the effects on pH of doubling or halving any initial values of P_{CO_2} or $[HCO_3]$ – or, as in question 7.2.4, quickly assess other comparable changes. This has a specific usefulness in graph-drawing, as will now be discussed.

A particularly useful form of graph for illustrating basic principles is one of plasma bicarbonate concentration against P_{CO_2}. Fig. 7.1 employs these as axes and shows one point to represent normal values in arterial plasma. A line is drawn to pass through this point and the origin. This corresponds to a constant ratio of bicarbonate concentration to P_{CO_2} and hence to a constant pH (equation 7.1). Whatever the values of pK_1' and S are, one knows that this pH has to be the normal 7.4, since the two other variables are normal. Two other lines may be added (Fig. 7.2) corresponding to pH 7.7 (i.e. 7.4 + log 2) and pH 7.1 (i.e. 7.4 − log 2). The first is obtained, as illustrated in Fig. 7.2, by plotting a point either for normal P_{CO_2} and twice-normal bicarbonate concentration (A), or for half-normal P_{CO_2} and normal bicarbonate concentration (B). The line for pH 7.1 is obtained in similar manner. How this 'scaffolding' may be used is illustrated in Fig. 7.3, which shows changes in plasma bicarbonate in response to metabolic acidosis and its respiratory compensation.

In the preparation of Figs 7.1 to 7.3 the values of S and pK_1' were not required. In other contexts it can help to combine them as (pK_1' − log S).

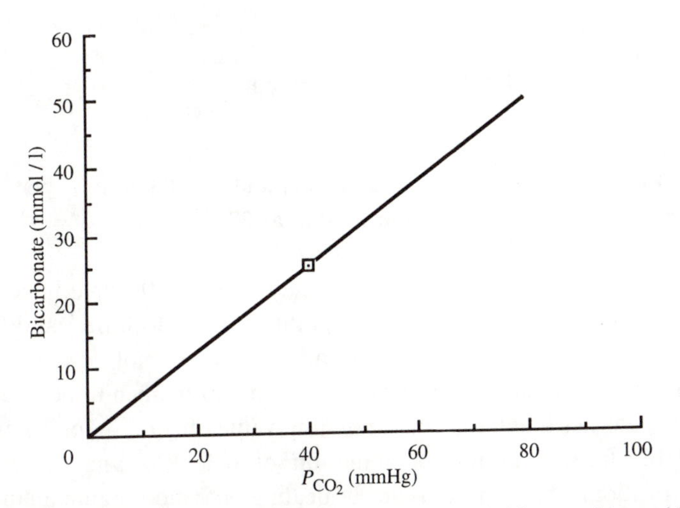

Fig. 7.1. Illustrating the combination of axes recommended for graphing acid–base changes. The single plotted point is for normal arterial plasma. The line drawn through that point shows those combinations of P_{CO_2} and bicarbonate concentration that correspond to the normal pH of 7.4.

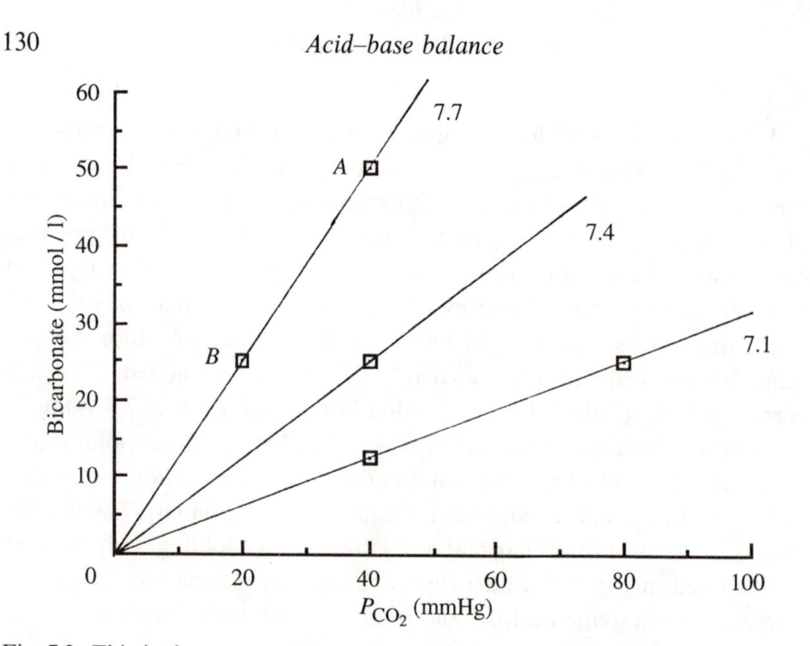

Fig. 7.2. This is the same as Fig. 7.1, except that two more lines of constant pH have been added. The line for pH 7.7 passes through points *A* and *B*. *A* represents a twice-normal bicarbonate concentration and normal P_{CO_2}. *B* represents a normal bicarbonate concentration and half-normal P_{CO_2}. The line for pH 7.1 is obtained in a similar way.

Then,

$$\text{pH} = (\text{pK}_1' - \log S) + \log \frac{[\text{HCO}_3]}{P_{CO_2}}. \tag{7.4}$$

7.2.4 **From the above values of pK_1' (6.1) and S (0.03 mmol/l per mmHg), what is ($\text{pK}_1' - \log S$) in plasma at 37 °C?** (log 0.03 = −1.52)

Combining the constants in this way may save a little time in repeated calculations, but that is not the only point to note. Both pK_1' and S vary with temperature and salt concentration and are not always readily available for particular conditions that one might happen to be interested in. On the other hand, the corresponding value of ($\text{pK}_1' - \log S$) may be calculable from a known combination of pH, P_{CO_2} and bicarbonate concentration. This can be useful in dealing with non-mammalian body fluids, but the question that follows refers to human plasma. It is provided for practice, but the answer may be obvious without further calculation, from what has gone before.

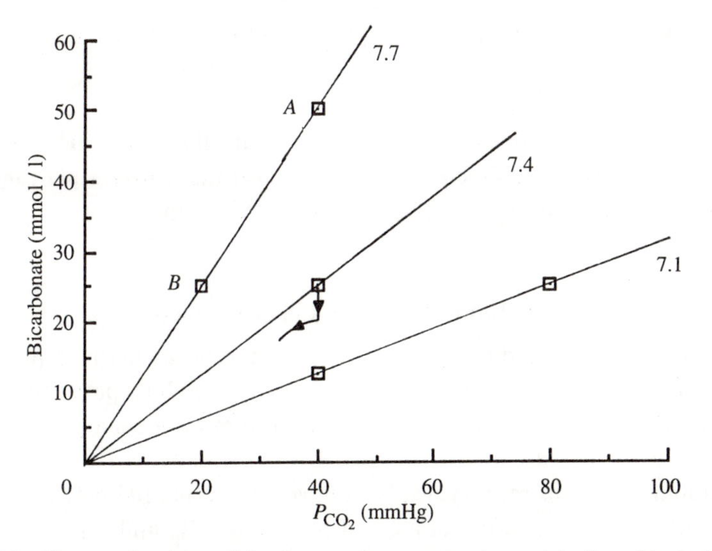

Fig. 7.3. Changes in plasma bicarbonate in response to metabolic acidosis. This illustrates the use of the 'scaffolding' of Fig. 7.2, applying it to the particular example of metabolic acidosis. Starting from the point representing normal acid–base balance, the vertical arrow represents the decrease in bicarbonate concentration that constitutes the metabolic acidosis itself; the curve to its left shows the progressive changes in P_{CO_2}, bicarbonate and pH that are associated with respiratory compensation.

7.2.5 **For comparison with the answer to 7.2.4, what is (pK_1' − log *S*), given that plasma at 37 °C has a pH of 7.4 when P_{CO_2} is 40 mmHg and [HCO_3^-] is 24 mmol/l?** (log 0.6 = −0.22)

The Henderson–Hasselbalch equation is commonly used in calculating bicarbonate concentrations from known values of pH and P_{CO_2}. How accurately should the constants (or the combined constant) be known or specified?

7.2.6 **If a bicarbonate concentration were calculated from pH and P_{CO_2}, by what percentage would it be in error if the assumed pK_1' (or else the measured pH) were out by 0.01 unit?** (antilog 0.01 = 1.023)

The relevant equilibria in plasma and other fluids are more complicated than is usually stated. One reflection of this is that pK_1' decreases somewhat with increasing pH. The exact change in pK_1' per unit change in pH depends on the circumstances (and the experimenter, it seems) but

pK'_1 has typically been found to fall by about 0.05 units for a rise in pH of 1 unit (i.e. $(\Delta pK'_1/\Delta pH) = -0.05$).

7.2.7 **In the calculation of bicarbonate concentrations from pH and P_{CO_2}, what percentage difference does this effect make over a pH range of 0.2 units?** (For a quick answer, see the previous question!)

7.3 Intracellular pH and bicarbonate

Are hydrogen ions and bicarbonate ions at equilibrium across typical cell membranes, or do they have a net tendency to diffuse passively either inwards or outwards? The answer is central to our understanding of how cell pH is regulated.

Mammalian cells seem typically to have an internal pH between about 6.8 and 7.4. The usual pH of arterial plasma is 7.40, and that of venous plasma 7.36. Interstitial fluid may be slightly more acid than plasma. Thus the pH inside a cell may be similar to the pH outside, but is usually slightly lower. One's first thought might therefore be that hydrogen ions tend generally to diffuse outwards. However, ionic diffusion is affected by electrical gradients as well as by concentration gradients, and the usual internal negativity of a cell must tend to draw hydrogen ions inwards. To find the net effect of these two opposing forces we may apply the Nernst equation, as we have done previously for other ions (equations 6.3, 6.13, 6.14, 6.17).

The equilibrium potential for hydrogen ions, E_H, is given, in millivolts at 37 °C, by the following equation:

$$E_H = 61.5 \log\{(H)_e/(H)_i\}. \qquad (7.5)$$

Note the use of activity here, i.e. (H), rather than concentration as when the Nernst equation was used before. This is actually more proper. Since (H) is defined here as 10^{-pH} (Section 7.1), equation (7.5) may be re-written in the following form that obviates the need to work with logarithms:

$$E_H = 61.5\{pH_i - pH_e\}. \qquad (7.6)$$

For purposes of calculation, here are some definite figures to work with:

$$pH_e = 7.4,$$

$$pH_i = 7.0,$$

$$\text{membrane potential, } V_m = -70 \text{ mV}.$$

7.3.1 **What is the equilibrium potential for hydrogen ions, E_H?**

7.3.2 **What is the electrochemical potential difference across the cell membrane, given by $(E_H - V_m)$?**

What is implied by these answers about the direction of net diffusion of hydrogen ions? (It may help to think of the analogy of sodium – Section 6.10.) If you are unsure how to relate signs and directions, try the following approach.

7.3.3 **For the same values of pH_e (7.4) and V_m (−70 mV), what would the value of pH_i have to be for hydrogen ions to be in equilibrium across the cell membrane (i.e. with $E_H = V_m$, and V_m substituted for E_H in equation 7.6)?** (70/61.5 = 1.14)

Since the actual pH is higher than this, it must be that hydrogen ions have a greater tendency to diffuse passively into the cell than out.

We may ask the same questions about bicarbonate ions, but take a short-cut to the conclusion. The Henderson–Hasselbalch equation (equation 7.1) relates pH, P_{CO_2} and $[HCO_3]$. Let us assume that the values of P_{CO_2} and pK'_1 are virtually the same inside and outside the cell. Then,

$$pK'_1 - \log(S \cdot P_{CO_2}) = \log[HCO_3]_i - pH_i$$
$$= \log[HCO_3]_e - pH_e. \quad (7.7)$$

Since $pH = -\log(H)$,

$$[HCO_3]_i \cdot (H)_i = [HCO_3]_e \cdot (H)_e,$$

or

$$[HCO_3]_i/[HCO_3]_e = (H)_e/(H)_i. \quad (7.8)$$

We may note in passing that equation (7.8) is suggestive of the Donnan equilibrium in its form (see equation (6.2)) but clearly it has a different basis. What equation (7.8) actually shows is that bicarbonate and hydrogen ions are quantitatively in an identical state of disequilibrium, except that the directions of net diffusion are opposite – hydrogen ions tending to enter the cell and bicarbonate ions tending to leave. If the basis of this conclusion is unclear, take the logarithm of each side of the equation and multiply by 61.5; the equation then shows that the equilibrium potentials for bicarbonate and hydrogen ions are equal (see equation (7.5)).

Consistent with this conclusion is the fact that there are mechanisms of membrane transport that act in directions that oppose these tendencies to passive diffusion. These include Na/H exchange, with sodium entering down its electrochemical gradient and thereby supplying the energy to drive hydrogen ions out of the cell, and also what is effectively a counter-transport of HCl and $NaHCO_3$, again driven by sodium entry.

We should not leave this topic without quantifying $[HCO_3]_i$. For this purpose we may use equation (7.7), according to which $(pH_e - pH_i)$ is equal to $(\log[HCO_3]_e - \log[HCO_3]_i)$ and so to $\log([HCO_3]_e/[HCO_3]_e)$.

7.3.4 **If the difference between pH_e and pH_i is 0.3 units (e.g. 7.4 minus 7.1) and $[HCO_3]_e$ is 26 mmol/kg water, what is $[HCO_3]_i$?**
(antilog 0.3 = 2)

This calculation yields merely a representative value for $[HCO_3]_i$. For the same value of $[HCO_3]_e$, and a range of intracellular pH of 6.8–7.4, $[HCO_3]_i$ would be 6.5–26 mmol/kg water. That $[HCO_3]_e$ also varies around the chosen value suggests an even greater variability in $[HCO_3]_i$.

For those who are content at this stage to think in terms of round numbers, the conclusion from the following calculation might be appreciated as an *aide mémoire*. The answer is used later.

7.3.5 **If $[HCO_3]_i$ is taken as typically half of $[HCO_3]_e$, and if the mass (or volume) of intracellular water in the whole body is taken as twice the mass of extracellular water, what is the total amount of intracellular bicarbonate in the body divided by the total amount of extracellular bicarbonate?**

7.4 Why bicarbonate concentration does *not* vary with P_{CO_2} in simple solutions lacking non-bicarbonate buffer

When the carbon dioxide tension rises in a solution containing bicarbonate and a non-bicarbonate buffer (e.g. protein), the concentration of bicarbonate rises too (Fig. 7.4). The mechanism is easily understood (see below), but what puzzles some people is the fact that concentration of bicarbonate varies hardly at all with carbon dioxide tension (over the physiological range) when the solution contains no non-bicarbonate buffer. The aim here is to clarify that point.

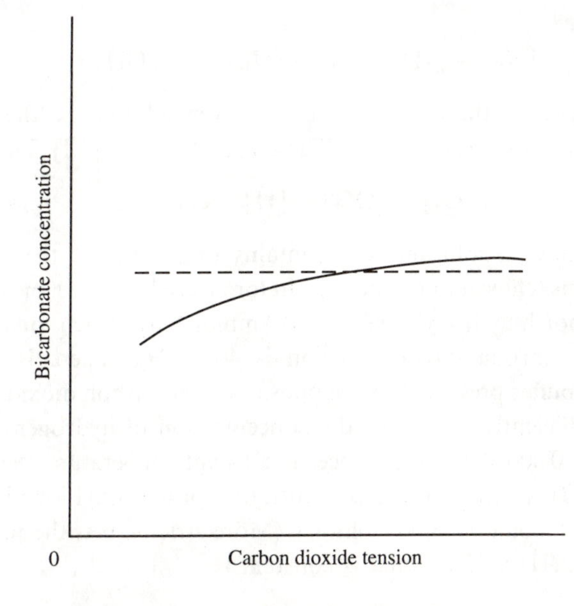

Fig. 7.4. Relationships between bicarbonate concentration and carbon dioxide tension in two solutions. One, like blood plasma, contains non-bicarbonate buffers such as proteins (solid line). The other contains no non-bicarbonate buffer (broken line).

Carbon dioxide reacts with water to produce bicarbonate and hydrogen ions:

$$CO_2 + H_2O = H_2CO_3 = H^+ + HCO_3^-. \qquad [7.2]$$

When there is non-bicarbonate buffer present (represented here as B), the hydrogen ions are mostly buffered as follows:

$$H^+ + B = BH^+. \qquad [7.3]$$

Overall, the reaction is

$$CO_2 + H_2O + B = HCO_3^- + BH^+. \qquad [7.4]$$

It is clear from this that bicarbonate generation is linked virtually one-to-one with non-bicarbonate buffering. Therefore, it should not happen significantly in the absence of non-bicarbonate buffer.

And yet the doubt may remain; after all, a solution does become more acid when the carbon dioxide tension rises and this change in pH implies the occurrence of reaction [7.2].

Consider a solution containing sodium, chloride and bicarbonate ions, but no non-bicarbonate buffer. From the principle of electroneutrality

(Section 6.5),

$$[Na] + [H] = [Cl] + [HCO_3] + [OH]. \tag{7.9}$$

Two of the terms in this are constant for a given solution, i.e. they cannot vary with carbon dioxide tension. These are [Na] and [Cl]. Thus,

$$[HCO_3] + [OH] - [H] = \text{constant}. \tag{7.10}$$

Consider now a solution that contains these ions at the following concentrations (chosen to be about right for arterial plasma): bicarbonate, 24.00000 mmol/l; hydroxyl ions, 0.00010 mmol/l; hydrogen ions, 0.00004 mmol/l. The carbon dioxide tension is 40 mmHg. There is no non-bicarbonate buffer present. Next, suppose that the carbon dioxide tension is altered sufficiently to double the concentration of hydrogen ions (i.e. [H] becomes 0.00008 mmol/l). Since, at constant temperature, the product ($[H] \times [OH]$) is constant, the concentration of hydroxyl ions is halved. Thus, [OH] becomes 0.00005 mmol/l. (More strictly, it is the product of the activities, $(H) \times (OH)$, that is constant.)

7.4.1 **According to equation (7.10), what is the new concentration of bicarbonate? Is it measurably different from 24.00000 mmol/l?**

7.4.2 **What is the final carbon dioxide tension?**

To some this may seem pernickety book-keeping; a man with a million pounds may spend a penny to significant effect, yet clearly leave undiminished his status of millionaire.

One anion has been ignored here that must be present with the bicarbonate. This is the carbonate ion. It has been ignored mainly for simplicity, but also because it is not generally what the aforesaid puzzled people have in mind. We look at carbonate next.

7.5 Carbonate ions in body fluids

The carbonate ion rarely features in elementary accounts of mammalian acid–base balance. Following on from the previous section, it is discussed here mainly in relation to the dependence of bicarbonate concentration on carbon dioxide tension. However, carbonate is also of interest to physiologists because of its presence in bone and in the shells of birds' eggs and of molluscs and crabs (where it occurs as calcium carbonate).

To those experimenters that prepare physiological salines, carbonate can be a nuisance if its concentration is allowed to rise so high that it precipitates with calcium.

Carbonate forms from bicarbonate thus:

$$HCO_3^- = CO_3^{2-} + H^+. \qquad [7.5]$$

At pH 7.4, the concentration of carbonate in blood plasma is about 0.1 mmol/l. That should also be true of the plasma-like solution described in the previous section, though, for simplicity, this was ignored. With the postulated doubling of P_{CO_2} and consequent lowering of pH to 7.1, the concentration of carbonate would be lowered, but, as with hydrogen and hydroxyl ions, its concentration starts low enough that only tiny amounts of extra bicarbonate could be formed.

With substantially falling P_{CO_2}, hence rising pH, it is a different matter; then, at tensions very much below normal, the concentration of bicarbonate declines markedly. As bicarbonate ions dissociate (i.e. reaction [7.5] proceeds to the right), the released hydrogen ions tend to decompose other bicarbonate ions in the following overall reaction:

$$2HCO_3^- = CO_3^{2-} + CO_2 + H_2O. \qquad [7.6]$$

The relationship between carbonate and bicarbonate is given by the following equation, in which pK_2' is the dissociation constant for reaction [7.5] (being, in other words, 'the second dissociation constant of carbonic acid').

$$\log \frac{[CO_3]}{[HCO_3]} = pH - pK_2'. \qquad (7.11)$$

7.5.1 **If pK_2' is 9.8, what is the ratio of carbonate to bicarbonate at (a) pH 7.8, (b) pH 8.8, (c) pH 9.8?**

Consider again a solution containing 24 mmol/l bicarbonate and no non-bicarbonate buffer. Its P_{CO_2} is 40 mmHg and its pH is 7.4. Now suppose that carbon dioxide is lost until the pH rises to 7.8. In accordance with the last answer and with reaction [7.6], only about 2% of the bicarbonate is converted to carbonate. Since the pH rises above 7.7, and therefore by more than 0.3 units, the P_{CO_2} must evidently fall somewhat below half of the initial value of 40 mmHg. (Note, yet again, the use of log 2 to achieve a rough answer.) The true P_{CO_2} is actually about 16 mmHg.

Plasma does not become more alkaline than pH 7.8 in respiratory alkalosis, but let us consider the effects of further loss of carbon dioxide, as when blood is exposed to air. If the pH were to rise to 8.8, then the solution would contain 20.0 mmol/l bicarbonate and 2.0 mmol/l carbonate. The value of P_{CO_2}, calculated from the Henderson–Hasselbalch equation, would be only 1.3 mmHg. The next question is for readers wishing to check this value of P_{CO_2} for themselves. Equation (7.4) may be used, with ($pK_1' - \log S$) taken to be 7.62 (the value calculated in question 7.2.5).

7.5.2 **What is the pH when, as above, the P_{CO_2} is 1.3 mmHg and the bicarbonate concentration is 20 mmol/l?** (log 15.4 = 1.19)

To summarize, bicarbonate concentrations in solutions free of non-bicarbonate buffer do fall substantially when the carbon dioxide tension is reduced far enough. The pH is then well above even pathological values, so that the phenomenon is of no importance *in vivo*.

7.6 Buffering of lactic acid

Human blood plasma typically contains about 0.7–2.0 mmol/l lactate, but the concentration can rise as a result of anaerobic metabolism, as in exercise, and as a compensatory response to alkalosis.

Although lactate in the body fluids is often referred to as 'lactic acid', in reality it is nearly all ionized. The next question is for those who would like to explore this point before returning to the matter of buffering.

The ratio of lactic acid to lactate is given by the following expression, this being derived by simple rearrangement from the definition of the equilibrium constant, pK.

$$\log \frac{[\text{lactic acid}]}{[\text{lactate}]} = \text{pK} - \text{pH}. \qquad (7.12)$$

The pK is about 4.6 in human plasma.

7.6.1 **What is the ratio of lactic acid to lactate at (a) pH 6.6 and (b) pH 7.6?** (Most human body fluids are within this range of pH)

Consider now how much the pH of extracellular fluid falls on addition of a given amount of lactic acid. The hydrogen ions are buffered by bicarbonate (with release of carbon dioxide) and to a lesser extent by proteins. The inorganic phosphate of extracellular fluid contributes rather little to buffering, but further hydrogen ions are taken up by erythrocytes, and, in the long term, by other tissues too.

There is an obvious advantage in considering a simpler situation here, so let us suppose that 5 mmol/l lactic acid are added to a salt solution of pH 7.4 containing 25 mmol/l bicarbonate and no other buffers. Of that bicarbonate, 5 mmol/l are turned to carbon dioxide and water, but the carbon dioxide is then lost, we also postulate, with readjustments of the P_{CO_2} to its original value. The reaction is as follows:

$$CH_3CH(OH)COOH + HCO_3^- = CH_3CH(OH)COO^- + H_2O + CO_2. \qquad [7.6]$$

The point of interest now is the fall in pH. Since the P_{CO_2} returns to its original value, the one change that is relevant to calculating the new pH is the fall in bicarbonate concentration from 25 to 20 mmol/l. The Henderson–Hasselbalch equation could be used as it stands, i.e. equations (7.1) or (7.4), but it is easier to use equation (7.3). With P_{CO_2} constant, we have

$$\begin{aligned} pH_1 - pH_2 &= \log[HCO_3]_1 - \log[HCO_3]_2 \\ &= \log \frac{[HCO_3]_1}{[HCO_3]_2}. \end{aligned} \qquad (7.13)$$

7.6.2 **In the above circumstances, with the bicarbonate concentration lowered from 25 to 20 mmol/l, P_{CO_2} unchanged, and the initial pH being 7.4, what is the final pH?** ($\log(25/20) = 1 - 3\log 2 = 0.1$)

This calculation provides an estimate of the fall in pH *in vivo* when P_{CO_2} is held constant through adjustments in ventilation, but P_{CO_2} would actually be adjusted to a lower value that would reduce the acidaemia (respiratory compensation). The acidaemia would also be reduced by the buffering effect of plasma proteins and, more slowly, by buffering within cells (Section 7.7) and bone (Section 7.8). The calculated fall in pH is therefore an overestimate of the fall *in vivo*.

Suppose, now, that all the carbon dioxide released from bicarbonate by the lactic acid were to stay in solution – with a consequent rise in P_{CO_2}. The concentration of bicarbonate falls again from 25 mmol/l to 20 mmol/l, but this time the concentration of dissolved carbon dioxide rises by 5 mmol/l. Its initial concentration can be taken as 1.2 mmol/l – normal for arterial plasma, as are values for initial pH and bicarbonate concentration. Thus the concentration of dissolved carbon dioxide (and the partial pressure also) rises by a factor of 6.2/1.2 or, approximately, 5. This is a convenient approximation because log 5 = (1 − log 2) = 0.7.

7.6.3 **What would the final pH be this time?**

The Henderson–Hasselbalch equation may be used again, but it is simpler to start with the answer to the previous question and adjust it for the change in P_{CO_2} by applying equation (7.3).

The difference between the two answers illustrates the special nature of the HCO_3/CO_2 buffer system – the fact that one component can be regulated by adjustments in pulmonary ventilation.

7.7 The role of intracellular buffers in the regulation of extracellular pH

Acid–base disturbances may be respiratory or metabolic and may involve acidosis or alkalosis. For simplicity, we concentrate here on respiratory acidosis. When the mean body P_{CO_2} is increased, the resulting fall in extracellular pH is moderated by a variety of mechanisms that raise the bicarbonate concentration in the extracellular fluid. Important amongst these is buffering by plasma proteins, erythrocytes, muscle and other cells, and bone. In addition, bicarbonate is generated by the kidneys ('renal compensation') and, to a small extent, by reduction in the concentration of organic anions, mainly lactate. The role of the extrarenal mechanisms has been studied in nephrectomized mammals.

The rise in extracellular bicarbonate concentration in response to a given degree of respiratory acidosis varies, but a doubling of P_{CO_2} may lead – over an hour or so, and by extrarenal mechanisms – to a rise of about 3 mmol/l. The purpose of the next calculation is to quantify the usefulness of this in moderating the initial fall in pH. Remember that the fall in pH due directly to a doubling of P_{CO_2} is 0.3 unit (Section 7.2).

7.7.1 **For comparison with this fall in pH of 0.3 units, what is the effect on pH of raising the bicarbonate concentration from 25 to 28 mmol/l?**

(log 1.12 = 0.05)

One might suspect that this unimpressive effect is largely due to erythrocytes. If P_{CO_2} is raised from 40 to 80 mmHg in a sample of oxygenated blood *in vitro*, the plasma bicarbonate concentration rises by about

7 mmol/l (depending on the haematocrit and other factors). This is much more than the rise *in vivo*, but it must be remembered that the bicarbonate released from erythrocytes *in vivo* is shared with the interstitial fluid as well as with the plasma.

7.7.2 ***In vitro*, the concentration of bicarbonate in the plasma of a particular blood sample rises by 7 mmol/l when the P_{CO_2} is doubled. What would the corresponding rise be, if the same amount of bicarbonate left each erythrocyte, but (much as *in vivo*) became distributed in 15 l of extracellular fluid instead of in 3 l of plasma?**

If the actual rise *in vivo* is 3 mmol/l (see above), it would thus seem that erythrocytes are not responsible for all of it. This is true. However, it is not as easy as might seem to calculate (by difference) the contribution of the erythrocytes, for the situation *in vivo* is complicated. One reason is that the blood is partly oxygenated and partly deoxygenated, and this affects the buffering properties of the haemoglobin. Also, the quantity of bicarbonate released from the erythrocytes is influenced by the final concentration in the plasma and this is determined not only by the relative volumes, as in question 7.7.2, but also by the amounts of bicarbonate entering the extracellular fluid from elsewhere (see Notes and Answers). A better estimate of the contribution of erythrocytes to buffering requires more information, and more complicated calculations than are appropriate here.

The buffering of extracellular fluid by bone mineral and by cells other than erythrocytes ('tissue buffering') receives little or no mention in many accounts of acid–base physiology. Our calculations so far suggest why: the effect would seem to be small. There is actually much more to the phenomenon and, to see why, we must first consider the regulation of cell pH.

Much of the work on tissue buffering of extracellular fluid was carried out decades ago. Nowadays there is more interest in the study of pH regulation in single cells. To take again the case of raised P_{CO_2}, this initially makes the cytoplasm more acidic. The cell must then increase its content of bicarbonate if it is to recover its original pH. It may do this partly by intracellular buffering and partly by transport of ions between cytoplasm and extracellular fluid – bicarbonate inward or hydrogen ions outward. (These two processes are equivalent in effect, for both of them lead to an increase in the bicarbonate content of the cell and a loss of bicarbonate from the extracellular fluid.) A very small amount of bicarbonate may also be generated by metabolism of organic anions such as lactate.

So, to re-iterate, the cell with a raised P_{CO_2} achieves better pH regulation by increasing its bicarbonate content rather than by losing bicarbonate to the extracellular fluid. There is thus a potential conflict between the regulation of intracellular pH and the regulation of extracellular pH.

To emphasize this conflict between intracellular and extracellular pH regulation, let us consider a much simplified model of the body. In this model there are only two fluid compartments – extracellular and intracellular. There are no kidneys, bone, or erythrocytes and no non-bicarbonate buffers (e.g. proteins) in the extracellular fluid. The P_{CO_2} is the same in each compartment. The concentrations of organic anions do not change. There is no osmotic redistribution of water following the movement of ions between body compartments.

Let us make the extreme assumption that all of the cells are able to regulate their pH perfectly by membrane transport of hydrogen or bicarbonate ions. This has the convenient effect of making buffering irrelevant, as will now be explained. When P_{CO_2} is first raised, there would indeed be production of bicarbonate within the cells as a result of buffering (by proteins, dipeptides, phosphates, etc.) as in reaction [7.4], but as the intracellular pH returns to normal, the ionization of the intracellular buffers, which is pH-dependent, also returns to normal. It follows that the ultimate contribution of buffering is nil, and this means that the total amount of bicarbonate in the two body compartments, taken together, stays constant. As to the amounts of bicarbonate originally present in each of the two individual compartments, it suffices here simply to take them as equal. (For the basis of this assumption, see the answer to question 7.3.5.)

We may now explore the effect of a rise in P_{CO_2} on extracellular bicarbonate and pH. Although the calculations relate to a much-simplified model of the body, the qualitative conclusions will be clear and important.

7.7.3 **For cell pH not to change, by what factor would the intracellular bicarbonate have to rise if P_{CO_3} were to double?** (see equation 7.1)

7.7.4 **Given that the amounts of bicarbonate in the two compartments start equal, and that their total stays constant, how much bicarbonate would be left in the extracellular compartment after the doubling of P_{CO_2}?**

Worded in a way that is more appropriate to a real body, the implications of the calculation are as follows: if most of the cells are able to maintain a normal or nearly normal pH in the face of a raised P_{CO_2}, then the extracellular bicarbonate concentration and pH should fall substantially. This is not what happens in mammals, nephrectomized or not (though the effect has been observed in a species of salamander).

Of the many simplifying assumptions in the model, the one most significant in accounting for its unrealistic behaviour is almost certainly the assumption that the cells maintain perfect pH regulation. (The non-bicarbonate buffers of extracellular fluid, including plasma proteins, were excluded from the model, but are of little quantitative importance. The inclusion of erythrocytes would only make the model marginally more realistic.) Further analysis of what occurs in the real body in response to respiratory acidosis is beyond the scope of simple calculation, but what seems to happen is summarized in Fig. 7.5. Those cells best able to regulate their internal pH increase their bicarbonate content at the expense of the extracellular fluid. Other cells (including erythrocytes) produce bicarbonate by internal buffering and share some of it with the

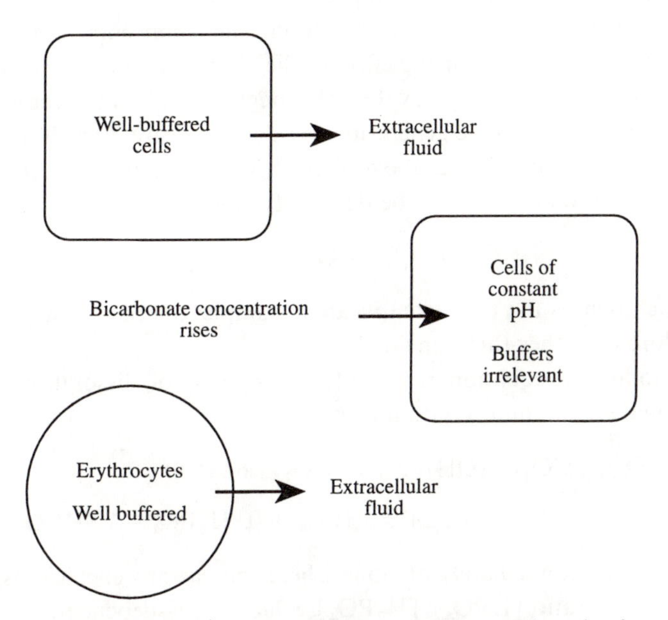

Fig. 7.5. Movements of bicarbonate ions (or of hydrogen ions in the opposite directions) between cells and extracellular fluid in response to raised P_{CO_2} (respiratory acidosis). The diagram represents a model organism in which some cells are able to maintain constant internal pH.

extracellular fluid. An uncertain amount of bicarbonate is also produced by buffering at the surface of bone mineral (see Section 7.8).

There is thus more movement of bicarbonate about the body than is implied by the small increase in plasma bicarbonate that is observed in the extrarenal response to respiratory acidosis. The redistribution of bicarbonate is all the more important because it is accompanied by movements of potassium and sodium in accordance with the principle of electroneutrality. Thus it is that the concentration of potassium in plasma tends to rise in acidosis. Changes in the distribution of sodium are less obvious, for reasons discussed in Section 6.3.

7.8 The role of bone mineral in acid–base balance

It is evident from the formulae of the various forms of bone mineral given in Section 6.4 that their precipitation and dissolution must affect acid–base balance. Moreover, it is common knowledge that bones may be dissolved in acid (for example, the bones of murder victims or bones in a dog's stomach), so it is obvious that dissolution must be accompanied by depletion of acid and by a rise in pH.

The extracellular fluid contains bicarbonate at a concentration of about 25 mmol/l and the alkalinizing effect of the dissolution of bone is to raise that concentration. We may explore the magnitude of that effect in terms of the ratio of bicarbonate generation to calcium release. For all the forms of bone mineral mentioned above that ratio is close to unity. The next paragraph shows how it may be derived for the case of

$$Ca_9(PO_4)_{4.5}(CO_3)_{1.5}(OH)_{1.5}.$$

This is chosen as the most complicated example; the reader may choose to explore the other three unaided.

The following equation represents the process of dissolution into a bicarbonate-containing solution at pH 7.4:

$$Ca_9(PO_4)_{4.5}(CO_3)_{1.5}(OH)_{1.5} + 8.4CO_2 + 6.9H_2O$$
$$= 9Ca + 3.6HPO_4 + 0.9H_2PO_4 + 9.9HCO_3 \quad [7.7]$$

It may not be immediately obvious where the various coefficients come from, but the ratio $[HPO_4]/[H_2PO_4]$, which is pH-dependent, needs to be set at 4 (i.e. 3.6/0.9, where 3.6 + 0.9 = 4.5) to be right for pH 7.4. (Concentrations of PO_4, and likewise CO_3 and OH, in extracellular fluid, are all small enough to ignore here, and this applies also to the concentration of non-bicarbonate buffer.) Equation [7.7] shows that 9.9 bicarbonate ions are produced for each 9 calcium ions released. For the purpose of answering the next question, the ratio may be taken as near enough 1.0.

7.8.1 **Suppose that the total concentration of calcium in the extracellular fluid is raised (very substantially) by 1 mmol/l through the dissolution of bone salts. What would be the final concentration of bicarbonate if the initial value were 25 mmol/l?** (Ignore again the small effects of non-bicarbonate buffers and treat the extracellular fluid as a closed system with respect to ions)

7.8.2 **What would be the change in pH if P_{CO_2} were maintained constant?** (log 1.04 = 0.017)

In interpreting this, note that, with care, plasma pH may be measured with capillary electrodes to a precision of about ±0.002. Routine measurements on other solutions with the usual, large, electrodes do not generally have an accuracy better than ±0.02.

These calculations seem to lead to a clear conclusion about the role of bone mineral in buffering – that it is small – but two important points need to be made in qualification. The first is that bone phosphates may have a significant role in the buffering of acid in the long term, provided that the dissolving calcium and phosphate are continually excreted so that plasma levels do not change. The second point is that bone mineral does not consist only of the above salts. Also present are sodium, potassium and what seems to be bicarbonate. It is likely that these are released in response to acidosis, as a separate and important mechanism of buffering.

7.9 The postprandial alkaline tide

For each hydrogen ion secreted into the stomach by the parietal cells, a bicarbonate ion is secreted into the blood. The tendency is therefore slightly towards metabolic alkalosis. Many textbooks refer to a consequent 'postprandial alkaline tide', a temporary reduction in the excretion of acid by the kidneys. This was discovered by Bence Jones in 1845. Few books give any quantitative information on the phenomenon.

Here we try to estimate the magnitude of the tide and of the associated change in plasma pH. For this we need to know how much acid is secreted after a typical meal, and how quickly. Amounts vary considerably from person to person and from meal to meal, with maximal rates varying from 1 to 47 mequiv/hour (see Notes and Answers). Let us settle for a modest 28 mequiv over 90 minutes (chosen to ease the arithmetic).

To calculate the effect on blood pH accurately, we would need to take into account the buffering that occurs in the plasma and other extracellular fluid, in the erythrocytes and other cells, and in bone. We also need to consider renal compensation and the time courses of all of these. The exercise is clearly too complicated for the back of an envelope, so let us start with an easier calculation that is based on gross over-simplification.

Accordingly, let us suppose that an individual secretes 28 mequiv of gastric acid, that all of this comes from 14 l of extracellular fluid, and that erythrocytes and the rest of the body make no contribution whatsoever to buffering and the adjustment of extracellular pH. (By excluding these other mechanisms we overestimate the effect on pH.)

The extracellular fluid initially contains, let us assume, 25 mmol/l bicarbonate and no non-bicarbonate buffer, so that the loss of acid means an equivalent gain of bicarbonate. Secretion of gastric juice implies loss of extracellular fluid, but let us ignore that.

7.9.1 **In this simplified situation, what is the final concentration of bicarbonate after secretion of the acid?**

7.9.2 **If there were no respiratory compensation (beyond removal of the excess CO_2 produced by the decomposition of bicarbonate) and the P_{CO_2} were thus to stay constant, by how much would the pH of the extracellular fluid rise?** (log 1.08 = 0.033)

The rise is small, but detectable. It might be less in a real body because (1) P_{CO_2} would tend to rise in compensation, and (2) some or all of the other acid–base mechanisms would come into play to lower the extra-cellular bicarbonate concentration. On the other hand, the quantity of acid secreted could be greater.

Considering now what might happen to renal excretion, let us start by utilizing an easily found item of quantitative information: typical rates of acid excretion over 24 hours are about 40–110 mequiv/day. These figures represent the sum of titratable acid and ammonia. (Note that this acid is also associated with feeding, inasmuch as it reflects such aspects of food composition as sulphur content.) Over shorter periods, the range of variation is much greater and it includes the negative values that correspond to the excretion of alkaline urine.

7.9.3 **At these rates of acid excretion (40–110 mequiv/day), how much acid would be excreted over a 90 minute period (i.e. the same duration as was chosen for the gastric acid secretion)?**

A comparison of the 28 mequiv of secreted HCl with this answer suggests two conclusions. First, it is entirely credible that the kidneys should be able to moderate significantly the rise in plasma pH that is due to gastric acid secretion. Second, if much of the 28 mequiv of base added to the extracellular fluid were to be removed by the kidneys, the effect on the urine should be distinctly perceptible as an alkaline tide. However, several investigators have sought the alkaline tide without finding it. On the heels of gastric acid secretion comes the pancreatic secretion of bicarbonate and this must surely diminish the tide. It seems appropriate that the two processes should cancel out, more or less, both in the gastrointestinal tract and in the extracellular fluid. The generality of the postprandial alkaline tide remains unclear.

8

Nerve and muscle

Here is a field of physiology that is full of quantitation and mathematics, but opportunities for applying only shop-keeper's arithmetic are fewer than might be expected. In Sections 8.1 and 8.2 we look at conduction in nerve and in cardiac Purkinje fibres. Section 8.3 takes us a tiny step in the direction of integrated neuromuscular activity, but also has the humbler aim of supplying a context in which to think about time scales and the speeds of neuromuscular events. Section 8.4 relates muscle mass to the activities of chinning the bar and jumping in a simple, rough-and-ready way that avoids calculating forces on levers. The approach leads naturally to a consideration of phosphocreatine usage in muscle contraction. Finally, Section 8.6 integrates information on sarcomere dimensions, myofilament spacing and calcium concentration in the context of muscle activation.

Resting membrane potentials and action potentials have already appeared in Sections 6.5, 6.10 and 6.11. Tensions in arteriolar smooth muscle were considered in Section 3.5.

One may memorize the numerical dimensions of skeletal muscle fibres, yet not perceive that some of these fibres have the lengths and widths of human hair. A class of students beginning physics was asked to estimate the height of the Empire State Building, which could be seen from the window (1250 ft without its television aerial). Estimates ranged from 50 ft to 1 mile. How much harder it must be to feel comfortable with microscopic dimensions or milliseconds! If a stereocilium in the organ of Corti were scaled up to the size of the Empire State Building, how much would it sway in response to a whisper? Many other such problems hover on the edge of arithmetic and a handful is addressed below.

8.1 Myelinated axons – saltatory conduction

Myelinated axons vary in conduction velocity from a few m/s to 120 m/s. Consider a peripheral nerve fibre with a conduction velocity of 100 m/s and with internode lengths of 1.5 mm, values appropriate to a motor fibre with a diameter of about 18 μm.

8.1.1 **How long does it take an action potential to spread from one node of Ranvier to the next?**

This answer may be compared with the duration of a single action potential at a node. Figures in typical elementary textbooks depict the positive phase of a nerve action potential as lasting about 0.5–1 ms, with the peak occurring at 0.2–0.6 ms. These durations are all more than ten times the answer to question 8.1.1. What does this mean? The point here is that some of those books give the clear impression (which many others do not contradict) that saltatory conduction in myelinated nerve fibres involves a 'jump' from one node to the next, with a pause for regeneration at each before activation of the next node. Perhaps the word 'saltatory' (defined as leaping) seems to imply this too, though that was never the intention. What the calculation implies, and what has long been known, is that the action potential constitutes a wave of activity that spans many nodes simultaneously.

The above myelinated nerve fibre has an internode length of 1.5 mm and a diameter of about 18 μm. Most myelinated fibres are narrower than this, but have about the same ratio of internode length to diameter. It is important to visualize these relative dimensions, and casual microscopy may not help much, for, examined with a ×40 microscope objective, many Schwann cells are far longer than the field diameter. Fig. 8.1 shows two examples of myelinated fibres as depicted diagrammatically in textbooks.

8.1.2 **If the above myelinated fibre is drawn to scale with a diameter of 18 mm, what length of paper would be required to show two nodes of Ranvier?**

Conducted at 100 m/s, an action potential in the same nerve fibre would travel 1 m in 10 ms. In the case of a very narrow myelinated fibre with a diameter of, say, 2 μm, the conduction velocity could be about 10 m/s.

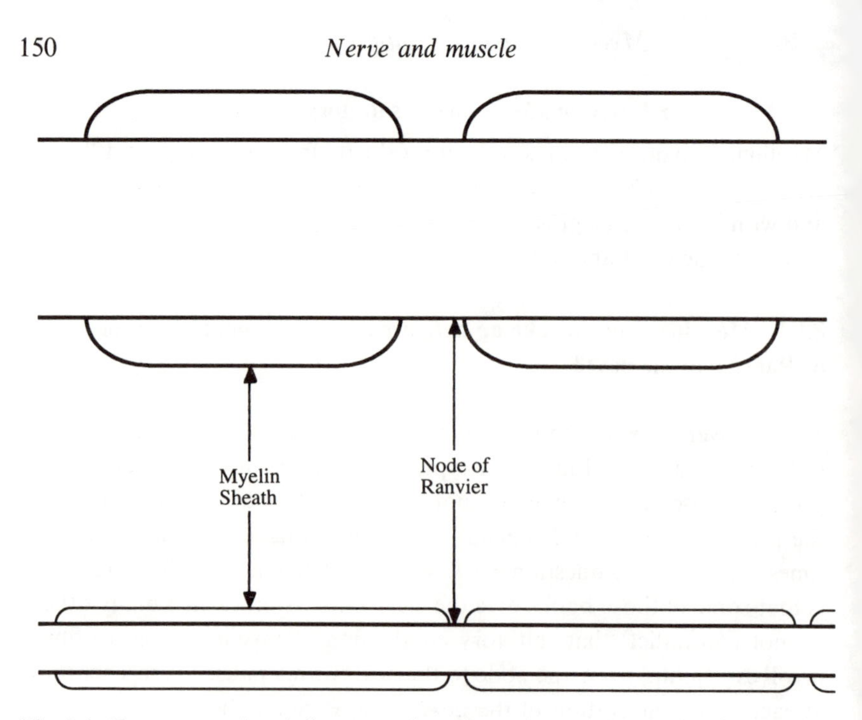

Fig. 8.1. Two conventional diagrams of myelinated nerve fibres, showing myelin sheaths and nodes of Ranvier.

8.1.3 **What difference, in ms, would this roughly tenfold reduction in diameter make to the time required for a motor response involving motor fibres (e.g. to muscles in the foot) that are 1 m in length?**

8.2 Non-myelinated fibres

C fibres are non-myelinated. The fastest conduct at about 2.3m/s and have diameters of about 1.1 μm. While conduction velocity is roughly proportional to fibre diameter in myelinated fibres, it is proportional to the square root of fibre diameter in non-myelinated fibres.

From this information, with the diameter in μm and velocity in m/s,

$$\text{velocity} = (2.3/\sqrt{1.1})\sqrt{(\text{diameter})} = 2.2\sqrt{(\text{diameter})}, \qquad (8.1)$$

or

$$\text{diameter} = (\text{velocity})^2/4.8. \qquad (8.2)$$

8.2.1 **The widest myelinated fibres conduct at about 120 m/s. Assuming the correctness of equation (8.2) for any diameter, how wide would a non-myelinated fibre have to be to have the same conduction velocity?**

Extrapolating like this is almost always unwise, but the general conclusion is safe: without myelination, our motor nerve fibres (and the nerves in which they run) would have to be very thick to conduct as rapidly as they need to.

Purkinje fibres in the heart need to be wide in order to conduct quickly enough for a rapid and properly coordinated heartbeat. With reservations, equation (8.1) may be tried on them as a test of its generality.

8.2.2 **A Purkinje fibre of diameter 100 μm may conduct at about 4 m/s. What is its conduction velocity calculated from equation (8.1)?**

This is hardly excellent agreement, but it is not as bad as might be anticipated, considering the differences between the action potentials in nerve fibres and Purkinje fibres, and the great extrapolation involved.

8.3 Musical interlude – a feel for time

We all have a feeling for hours, minutes and seconds. The milliseconds of physiology may need more thinking about to become meaningful. A vibration of 1000 cycles/s (i.e. 1 cycle/ms) in the vocal cords gives a musical pitch about one octave above middle C. The note from a 32-foot organ pipe may throb with a frequency near 16 cycle/s, about as low as most people are able to perceive as a note rather than as separate sounds (which would be spaced at about 60 ms).

A book of transcribed solos of the jazz saxophonist Charlie Parker shows that he improvised on the tune 'Bird gets the worm' at a rate of 340 beats per minute (this being the fastest rate in the collection). Mostly he was playing two notes (quavers) per beat, though sometimes three or four. Playing quavers he was thus fingering notes at a rate of 11/s.

8.3.1 **At that rate, how many ms were available for each note?**

Saxophones and other musical instruments can be played faster, especially when there is no improvisation involved and the fingers follow well-practised patterns of movement. With a frequency of, say, 13 notes/s (1 per 77 ms), one even approaches the throb rate of a deep organ pipe. For comparison, note that the syllables of human speech can barely exceed ten per second (though the motor commands needed for speech must number hundreds per second). As there is a similar limitation to mental counting, it is not actually very easy to judge the pace of the fastest music.

Part of the point of the above calculation is that it may give one a better feel for time intervals close to 100 ms (0.1 s), and a context in which to consider the time scales of simpler neural and neuromuscular phenomena. Here are some examples, although it is best to think about the matter in the context of one's own knowledge and interests. In a (long) myelinated axon with a (slow) conduction velocity of 10 m/s, it takes an action potential 100 ms to travel 1 m. The time elapsing between the start of an isometric contraction and the moment of peak tension in various human hand muscles after a single electrical stimulus to the nerve is 55–80 ms, while calf muscles may take twice as long to achieve peak tension. About 20 ms elapse between stimulation of the motor cortex and resulting action potentials in the finger muscles. Eye closure during a blink takes about 50–150 ms, and re-opening about 100–200 ms. Elite sprinters may take 4–5 strides per second, at 200–250 ms/stride. Humming birds beat their wings at 30–50 times per second, so that a single complete beat may take as little as 20 ms.

To return to the subject of music, even the most relevant of those examples is no explanation for the limits to fast instrumental playing. It is not the case that the fingering of each note is separated in time from the fingering of the next, for the motor activities overlap. This is less obviously the case with a trill, since this typically involves a single finger moving rapidly up and down. Depending on the finger and the person, the rate of movement could be 6–9/s. Since each movement in a trill produces two notes, the time for each note is 56–83 ms.

With regard to Charlie Parker, there is one particular question to be addressed: to what extent was there time for him to hear and react to one note before he played the next? A player can certainly do this in much slower music.

Consider 'reaction time' in the sort of experiment in which a subject presses a button in response to a sensory stimulus. If the latter is a sudden light, then the reaction time could be perhaps 150–250 ms, depending very much on the individual and circumstances. Reactions to sound, more relevant here, are usually about 30–50 ms faster. The reactions of elite sprinters to a starting gun can be as quick as those two ranges suggest. Reactions requiring a decision, such as which of several buttons to press, take substantially longer.

8.3.2 **Which gives the longer interval, the answer to question 8.3.1 or the reaction time to an auditory stimulus?**

A jazz musician is continually responding to the notes that he and the other musicians produce, but in fast playing there is clearly no possibility of useful feedback between successive notes. Rather, the fingers are responding to pre-programmed patterns of neural activity that correspond to groups of notes. How this is achieved, and without any feeling of delay, is hardly something to solve by simple arithmetic.

8.4 Muscular work – chinning the bar, saltatory bushbabies

The mechanics of human movement would seem to offer considerable scope for quantitative treatment involving muscle tensions and the arithmetic of levers. However, a simple limb movement may call on a number of muscles working together, and the relevant measurements may be hard to elicit, even from many shelves of anatomy books. Here we side-step these complexities to utilize an idea of seductive simplicity: the work of which a skeletal muscle is capable, in a single contraction *in situ*, is roughly proportional to its mass or volume. The reasoning behind that is as follows.

The forces that skeletal muscles exert are proportional to their cross-sectional areas. The distances that they are able to shorten *in situ* are proportional to their resting lengths (as well as depending on their anatomical location). The work that they can do in a single contraction is given by the force multiplied by the shortening distance, and is therefore proportional to volume. Muscle volume is proportional to muscle mass.

In their normal movements the locomotory skeletal muscles are said to be able to shorten, typically, by up to 25%. (Some put the figure at 30%, but muscles vary in their ability to shorten and it is hard even to define an average.) Because of this limited shortening, the tensions that muscles are able to exert are close to the maxima of their length–tension curves. This means that, when shortening is isometric or very slow, stresses can be about $3\ \text{kg/cm}^2$, or $3 \times 10^5\ \text{N/m}^2$. Since the density of muscle is about $1060\ \text{kg/m}^3$, the work that can be expected of a typical muscle in a single slow, strong contraction is $3 \times 10^5 \times 25/100 \div 1060 = 70\ \text{J/kg}$. This value will be assumed in the following calculations, but it is just an estimated typical value. Moreover, the faster a muscle contracts, the more are the tension and work reduced. For example, the faster rate of contraction that is required to maximize the power output reduces the tension to 30% and this lowers the work output to $0.3 \times 70 = 21\ \text{J/kg}$.

Consider the 'chin-up', the activity of hanging from a bar by one's hands, then drawing oneself up to touch it with the chin. A reason for choosing this activity is that some individuals can do it easily and some hardly at all; for some people one can therefore suppose that the relevant muscles are exerting their maximum tensions. The work required to raise the body is calculated as force multiplied by distance. Raising 1 kg by 1 m requires 9.8 J (i.e. 1 kg·m = 9.8 J).

8.4.1 **During a chin-up, a particular individual's body is lifted 0.4 m. How much work is required *per kg of body mass*?**

8.4.2 **Assuming that this individual can only just accomplish this action – slowly – and that the relevant muscles perform work at 70 J/kg of muscle, what must the total mass of these muscles be as a percentage of body mass?**

For comparison, the total percentage of muscle in a standard man of 70 kg (containing 14% fat) is about 43%. Whether the answer to 8.4.2 is right for an individual just able to perform a chin-up is hard to check. The relevant muscles dissected from one cadaver (see Notes and Answers) were found to constitute only 1.4% of the body mass, but the man was 80 years old. Considerable amounts of muscle are lost in old age.

Actual body mass is not itself specified in these calculations, but the answers did depend on arm length. Consider now a smaller person with the same bodily proportions.

8.4.3 **This smaller person has arms half as long those of the previous individual. What percentage of muscle would be required to achieve a chin-up?**

There is no implication here concerning the actual muscularity of this person; the conclusion is rather that small boys should generally find it easier to lift themselves than their fathers do. Another route to this last conclusion is given in Notes and Answers.

In high jumping, the advantage is with the taller individual because the centre of gravity starts high, but if jumping is assessed in terms of the raising of the centre of gravity, then tall people have no advantage.

Terrestrial mammals of different sizes tend to jump to similar heights, but the information on athletes, horses, etc., rarely relates to centres of gravity. Two metres is regarded as a suitable height of fence to keep out antelopes, regardless of their size. High jumping as an athletic event involves a run and then a motion of the body that allows the centre of gravity to pass under the bar. Such complexities will be avoided here by considering a standing jump, that of the bushbaby (*Galago senegalensis*). A jump has been recorded for a bushbaby weighing of 0.25 kg in which the centre of gravity was raised about 7 ft, i.e. 2.1 m.

8.4.4 **A bushbaby leaps upwards, raising its centre of gravity by 2.1 m. How much mechanical work is required per kg of body mass?**
(1 kg·m = 9.8 J)

8.4.5 **On the basis of this answer, and the supposition (of distinctly questionable appropriateness here) that muscles typically achieve 70 J/kg in a single slow, strong contraction how much muscle, expressed as a percentage of body mass, is involved in the jump?**

Measurements made on two bushbabies showed that the muscles of the hind legs and back, not all used in jumping, made up 24–25% of the body mass. The total muscle content of the body was 36–37%. What does this mean? How appropriate were the various assumptions?

8.5 Phosphocreatine in muscular contraction

Energy for muscular contraction is provided by the hydrolysis of adenosine triphosphate (ATP). An additional reservoir of energy is provided by phosphocreatine and it is this, rather than the ATP, that is first depleted during muscular activity. The link between phosphocreatine and ATP is through the Lohmann reaction:

$$\text{phosphocreatine} + \text{ADP} = \text{creatine} + \text{ATP}. \qquad [8.1]$$

The concentration of phosphocreatine in resting skeletal muscle is commonly 15–25 mmol/kg. This is considerably higher than the concentration of ATP itself, which is nearer 5 mmol/kg. Here we relate the phosphocreatine content of skeletal muscle to work of contraction.

In Section 8.4 a rough value was derived for the amount of work that might be performed by a typical skeletal muscle *in vivo* in a single slow contraction, maximal in terms of tension and shortening. This is 70 J/kg. Since this is expressed per kilogramme, like the concentrations of muscle phosphocreatine and ATP, and since the 'energy contents' of these per mole are known, it is tempting to put the information together to see what emerges about the usage of phosphocreatine or ATP per contraction.

For example, in one contraction, might all the phosphocreatine be used up, hardly any, or some intermediate 'substantial proportion'?

There is no need to consider the (fairly small) free energy change of the Lohmann reaction. Since one molecule of phosphocreatine yields one molecule of ATP, it suffices to consider the free energy change in the hydrolysis of ATP to ADP, ΔG_{ATP}. For cellular conditions, this was given in Section 2.2 as -10 to -13 kcal/mol, or -42 to -54 kJ/mol. Taking it for convenience as -50 kJ/mol, or -50 J/mmol, we have a figure to compare directly with the mechanical work of contraction. In doing this, we may start by assuming complete conversion of chemical energy to external work, and only then give thought to the considerable loss of energy as heat.

8.5.1 **Assuming completely efficient energy conversion, if a single slow contraction, maximal in terms of tension and of *in vivo* shortening, accomplishes external work to the extent of 70 J/kg and the energy available from ATP is 50 J/mmol, by how much might the concentration of phosphocreatine fall?**

This answer may be compared with the resting concentration of phosphocreatine of roughly 15–25 mmol/kg muscle. Too much should not be made of the exact answer, however. The less efficient the actual energy conversion, the more phosphocreatine would actually be needed per contraction. There is no fixed figure for the percentage efficiency – as is particularly obvious if one thinks of isometric contraction against so great a load that no evident shortening occurs. Then, no mechanical work is done and the efficiency is zero. In contrast, the mechanical work done in a strong, rapid contraction against a smaller force is less than 70 J/kg muscle (Section 8.3), so that less phosphocreatine would be used.

Abandoning precise quantitation, one can at least say that single strong, slow contractions may use up phosphocreatine to the extent of 'at least several mmol/kg muscle'.

8.6 Calcium ions and protein filaments in skeletal muscle

Calcium ions play a crucial role in the activation of muscle. In the sarcoplasm of inactive skeletal muscle the concentration of free calcium ions is roughly 10^{-7} mol/l water. Maximum activation requires a rise in concentration to something like 2×10^{-5} mol/l water, at which concentration there is near-maximal binding of calcium to the molecules of troponin-C of the thin filaments. Since there is information readily available on the sizes of filaments and sarcomeres, on the spacing of filaments, and on the numbers of calcium-binding sites on the thin filaments, one may relate these to actual numbers of calcium ions. Some may find the results surprising. (In similar vein, the numbers of hydrogen ions in a mitochondrion were calculated in Section 7.1.)

A skeletal muscle fibre contains a large number of parallel, cylindrical myofibrils, each with a diameter of about 1–2 μm. Within the individual sarcomeres that constitute the myofibrils, the thick filaments are about 1.55 μm long and the thin filaments, anchored to the Z discs, are about 1 μm long. Since it is with the thin filaments that the calcium interacts, let us estimate, for the non-activated state, the number of free calcium ions in the portion of a half-sarcomere that contains thin filaments and is delimited by their ends.

8.6.1 **Suppose that the thin filaments at one end of a particular sarcomere are contained within a cylinder of length 1 μm and radius 0.5 μm. What are (a) the cross-sectional area of this cylinder in μm^2 and (b) the volume in μm^3?**

Although the cylinder is partially occupied by thick and thin filaments, it may be taken for present purposes as consisting only of water. (The error may be gauged roughly by looking at electron micrographs of muscle. It increases with the degree of overlap between the the thick and thin filaments.) To calculate the number of ions in this volume, recall that 1 mol of calcium, or any other substance, contains 6.0×10^{23} molecules (Avogadro's number).

8.6.2 **How many calcium ions are there in that 0.8 μm^3 of fluid, if the concentration is 10^{-7} mol/l?** ($1\ \mu m^3 = 10^{-15}$ l)

This answer may now be related to the number of thin filaments within this portion of the sarcomere. Seen in transverse section, the parallel thin filaments appear in hexagonal arrays (Fig. 8.2). Each thin filament lies in the middle of an equilateral triangle with thick filaments at the angles. The thick filaments are about 45 nm apart. From this information may be calculated first the number of thin filaments per unit area of cross section and then the number in the cylinder of sarcomere that was specified above.

The area of an equilateral triangle of side S is equal to $\sqrt{3}/4 \times S^2$

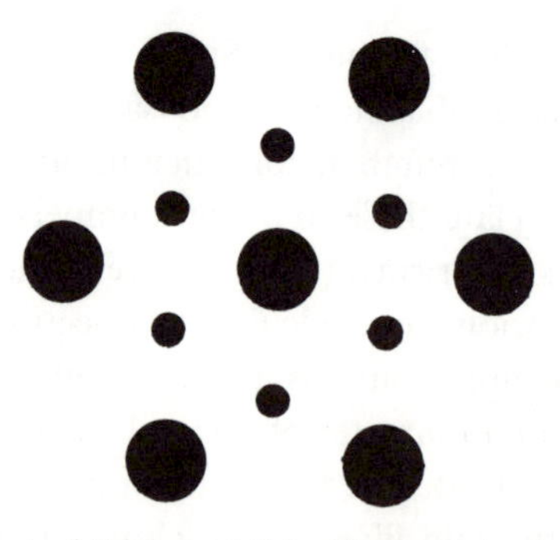

Fig. 8.2. The arrangement of thick and thin filaments in skeletal muscle as seen in transverse section.

(i.e. $0.43S^2$). The area containing 1 thin filament is thus $0.43 \times (45)^2$ nm^2 = 871 nm^2. This may be rounded to 0.9×10^{-3} μm^2.

8.6.3 **On the basis of this rounded figure, how many thin filaments are there in 1 μm^2 of cross section?** ($9 \times 11 \simeq 100$)

8.6.4 **How many thin filaments are there in our cylindrical portion of sarcomere with its cross sectional area of 0.8 μm^2**

In resting muscle this number of thin filaments shares the 0.8 μm^3 of sarcomere-end with about 48 calcium ions (calculated in 8.6.2).

8.6.5 **What is the ratio of thin filaments to calcium ions?**

The point of the series of calculations, apart from exercising quantities that may be already familiar, is simply the possible unexpectedness of the final answer. The intended conclusion might seem to be that the high ratio constitutes an explanation for the lack of tension in the sarcomere, but that is not so; the number of calcium ions bound by the troponin-C depends also on the affinity of one for the other and this is very high. (If that point is not clear, note that the mechanisms that maintain such low concentrations of calcium ions in the cytosol have also to be responsive to those same concentrations.)

During activation of the muscle the concentration of calcium ions rises perhaps 200-fold. This means that the number of free calcium ions increases from 48 to $48 \times 200 = 9600$ – let us call it 10 000. This lowers the ratio of thin filaments to free calcium ions to about 0.1. Even then, there is a large discrepancy between the number of free calcium ions and the number of molecules of troponin-C. These occur in pairs along the thin filament with a spacing of about 37 nm, implying that there are about $1000/37 = 27$ pairs on a 1 μm filament, or about 54 molecules. (This calculation disregards the disruption of the repeating pattern at the filament ends.) Since there are 880 thin filaments in our chosen sarcomere-end, there are about $880 \times 54 = 47\,520$ molecules of troponin-C.

On maximal activation, four calcium ions attach to each molecule of troponin-C. We have already estimated that the number of free calcium ions in the cylinder increases to about 10 000.

8.6.6 **What then is the ratio of bound calcium to free calcium?**

Evidently only a small part of the calcium entering this portion of the sarcomere on activation remains in free form, a much greater part becoming attached to troponin-C.

The thick filaments were taken as lying about 45 nm apart. Does this square with the appearance of myofibrils in electron micrographs? Seen in longitudinal section, the thick filaments span the sarcomeres in numbers that are definitely easily countable, i.e. in tens rather than in hundreds or thousands.

8.6.7 **In a myofibril of diameter 1.35 μm (chosen to be 3 × 450 nm), how many thick filaments should lie side-by-side across the width if their centres are 45 nm apart?**

Appendix A
A useful logarithm

Many students come to physiology still a little unsure about logarithms. Those who do understood them well may yet have had no cause to learn any numerical values, apart, that is, from the logarithms, to base 10, of 1, 10, 0.1, etc. (Table A.1). The major point to be made here is that it can really be worthwhile to remember at least one. With this one logarithm, one may calculate others, perform an unexpectedly large number of useful physiological calculations, and appear more of a mathematician than perhaps one is.

The recommended logarithm is $\log_{10} 2$. To five significant figures, it is memorable for its symmetry, being 0.30103, but for physiological calculations, 0.30 is usually close enough. This finds application a number of times in other parts of the book.

When a logarithm is needed, the odds of it being that particular one might seem to be slim. In a moment we will see how a few other logarithms may be quickly derived from it, but another point must be made first. It is this, that when one performs rough calculations in physiology, one is often free to choose the quantities involved and to choose round numbers and representative quantities. Here is an example relating to Chapter 7.

> Question: Could a rise in bicarbonate concentration from 20 mM to 30 mM explain a rise in pH from 7.10 to 7.43?
> Answer: No – even a doubling of concentration only leads to a rise of 0.3 unit at constant P_{CO_2}.

The remainder of this Appendix is about how to derive logarithms easily when one is parted from tables and calculator. Some people find aesthetic pleasure in such things; those who just find them stressful may yet care to proceed further, for the exercise.

9.1 **Computer users may recall that a kilobyte is not 1000 bytes, but 1024 bytes. Now, 1024 is 2^{10} and it is also nearly equal to 10^3. Taking 2^{10} as identical with 10^3, what is log 2?**

Armed with the logarithm of 2, let us now see how the logarithms of some other whole numbers may be obtained from it. To find log 4, note that 4 is equal to 2^2. It follows that log 4 is twice log 2 and is therefore calculable as $2 \times 0.301 = 0.602$.

Table A.1 *Some logarithmic relationships (logarithms to base 10)*

$\log 1 = 0$
$\log 10 = 1$
$\log 100 = 2$
$\log 0.1 = -1$
$\log(xy) = \log x + \log y$
$\log(x/y) = \log x - \log y$
$\log x^y = y \log x$

Table A.2 *Some logarithms (to base 10)*

x	$\log x$
1	**0**
1.6	*0.204*
2	**0.301**
2.5	*0.398*
3	0.477
3.5	*0.544*
4	**0.602**
5	**0.699**
6	0.778
6.25	*0.796*
7	**0.845**
8	**0.903**
9	0.954
10	**1.000**

Those values shown **bold** are those most simply obtained in the way described in the text. Those shown *italicized* are obtainable from these, making further use of log 2.

9.2 **Likewise, since $8 = 2^3$, $\log 8 = 3 \times \log 2$. What is log 8?**

9.3 **Since $5 = 10/2$, $\log 5 = (\log 10 - \log 2)$. What is log 5?**

9.4 **The square of 7 is very nearly half of 100. Log 7 is thus only slightly (0.5%) less than half of $(\log 100 - \log 2)$. What is that?**

Thus far we have the logarithms of 1, 2, 4, 5, 7, 8 and 10, and this is probably as far as one would usually wish to take the process in the context of physiological mental arithmetic. Table A.2 gives these values in bold type. It is easy enough to add others in the same way. Thus, x may be taken as 5/2, 7/2, $(8 \times 2/10)$, $10/(8 \times 2/10)$ and so on. These values are italicized in the table. Logarithms obtained in this way can be useful if one wishes to construct rough logarithmic graph paper without benefit of printed tables.

Those wishing to estimate log 3 by comparable means could start from the fact that 3^4 is close to 10×2^3 or that 3^9 is nearly 2×10^4. Doubling log 3, one obtains log 9. Log 3 plus log 2 gives log 6.

How accurate does a logarithm need to be? What is the implication of taking log 2 as 0.30 instead of 0.301? The antilogarithm of 0.300 is 1.9953 instead of 2 – a difference of 0.2%. In general, a difference in log x of 0.001 implies a difference in x of 0.23%. A difference in log x of 0.01 implies a difference in x of 2.3%.

References

Agostoni, E. & D'Angelo, E. (1991). Pleural liquid pressure. *Journal of Applied Physiology*, **71**, 393–403.

Alexander, R. McN. (1992). The work that muscles can do. *Nature*, **357**, 360–1.

Bangham, A. D. (1991). Pattle's bubbles and Von Neergaard's lung. *Medical Science Research*, **19**, 795–9.

Bowman, W. C. & Rand, M. J. (1980). *Textbook of Pharmacology*, 2nd edit. Oxford: Blackwell Scientific Publications.

Burton, A. C. (1965). *Physiology and Biophysics of the Circulation.* Chicago: Year Book Medical Publishers Inc.

Burton, R. F. (1965). Possible factors limiting the concentration of haemocyanin in the blood of the snail, *Helix pomatia* L. *Canadian Journal of Zoology*, **43**, 433–8.

Burton, R. F. (1973). The significance of ionic concentrations in the internal media of animals. *Biological Reviews*, **48**, 195–231.

Burton, R. F. (1987). On calculating concentrations of 'HCO_3' from pH and PCO_2. *Comparative Biochemistry and Physiology*, **87A**, 417–22.

Burton, R. F. (1988). The protein content of extracellular fluids and its relevance to the study of ionic regulation: net charge and colloid osmotic pressure. *Comparative Biochemistry and Physiology*, **90A**, 11–16.

Burton, R. F. (1992). The roles of intracellular buffers and bone mineral in the regulation of acid–base balance in mammals. *Comparative Biochemistry and Physiology*, **102A**, 425–32.

Clausen, T., Van Hardeveld, C. & Everts, M. E. (1991). Significance of cation transport in control of energy metabolism and thermogenesis. *Physiological Reviews*, **71**, 733–74.

Diem, K. (1962). *Documenta Geigy. Scientific Tables.* 6th edit. Manchester: Geigy Pharmaceutical Co. Ltd.

Driessens, F. C. M., Verbeeck, R. M. H. & van Dijk, J. W. E. (1989). Plasma calcium difference between man and vertebrates. *Comparative Biochemistry and Physiology*, **93A**, 651–4.

Durnin, J. V. G. A. & Passmore, R. (1967). *Energy, Work and Leisure*, Heinemann Educational Books Ltd.

Gil, J., Bachofen, H., Gehr, P. & Weibel, E. R. (1979). Alveolar volume–surface area relation in air- and saline-filled lungs fixed by vascular perfusion. *Journal of Applied Physiology*, **47**, 990–1001.

Guyton, A. C., Moffatt, D. S. & Thomas, H. A. (1984). Role of alveolar surface tension in transepithelial movement of fluid. In *Pulmonary Surfactant.* (eds B. Robertson, L. M. G. van Golde & J. J. Batenburg) pp. 171–85. Amsterdam: Elsevier.

Harvey, R. J. (1974). *The Kidneys and the Internal Environment.* London: Chapman and Hall.

Harwood, P. D. (1963). Therapeutic dosage in small and large mammals. *Science*, **139**, 684–5.

Hilden, T. & Svendsen, T. L. (1975). Electrolyte disturbances in beer drinkers. *Lancet*, **2**, 245–6.

Hofstadter, D. (1985). On number numbness. In *Metamagical Themas.* pp. 115–35. Harmondsworth: Viking, Penguin Books Ltd.

Kleiber, M. (1961). *The Fire of Life.* New York: John Wiley and Sons.

Lindstedt, S. L. & Calder, W. A. (1981). Body size, physiological time, and longevity of homeothermic animals. *Quarterly Review of Biology*, **56**, 1–16.

Mandelbrot, B. B. (1983). *The Fractal Geometry of Nature.* New York: Freeman.

Pfaller, W. & Rittinger, M. (1980). Quantitative morphology of the rat kidney. *International Journal of Biochemistry*, **12**, 17–22.

Pirofsky, B. (1953). The determination of blood viscosity in Man by a method based on Poiseuille's law. *Journal of Clinical Investigation*, **32**, 292–8.

Rahn, H. (1966). Aquatic gas exchange: theory. *Respiration Physiology*, **1**, 1–12.

Schafer, J. A. (1982). Salt and water absorption in the proximal tubule. *The Physiologist*, **25**, 95–103.

Schmidt-Nielsen, K. (1984). *Scaling: Why is Animal Size so Important?* Cambridge: Cambridge University Press.

Schwerzmann, K., Hoppeler, H., Kayar, S. R. & Weibel, E. R. (1989). Oxidative capacity of muscle and mitochondria: correlation of physiological, biochemical, and morphometric characteristics. *Proceedings of the National Academy of Sciences, USA*, **86**, 1583–7.

Stahl, W. R. (1965). Organ weights in primates and other mammals. *Science*, **150**, 1039–42.

Stone, H. O., Thompson, H. K. & Schmidt-Nielsen, K. (1968). Influence of erythrocytes on blood viscosity. *American Journal of Physiology*, **214**, 913–18.

Thaysen, J. H., Lassen, N. A. & Munck, O. (1961). Sodium transport and oxygen consumption in the mammalian kidney. *Nature*, **190**, 919–21.

Thurlbeck, W. M. & Wang, N-S. (1974). The structure of the lungs. In MTP International Review of Science, Physiology Series One, vol. 2, *Respiratory Physiology* (ed. J. G. Widdicombe), London: Butterworth and Baltimore: University Park Press.

West, L. J., Pierce, C. M. & Thomas, W. D. (1962). Lysergic acid diethylamide: its effect on a male Asiatic elephant. *Science*, **138**, 1100–2.

Appendix B

Notes and answers

In Chapter 1 I encouraged the reader both to make approximate calculations, where appropriate, and to write out all units in the calculation if there is a chance of confusion. Many of the answers are given here to a higher degree of precision than is necessary and units are not always given in the working.

1. Introduction to physiological calculation – approximation and units

The relationship between kidney mass and body mass is from Stahl (1965).

2. Energy and metabolism

Measures of energy

2.1.1 $100\ \text{kg} \times 0.8\ \text{kcal/kg} = 80\ \text{kcal}$.
2.1.2 (a) $9.8 \times 100 \times 10/1000 = 9.8\ \text{kJ}$,
(b) $9.8 \times 0.24 = 2.4\ \text{kcal}$.

Adenosine triphosphate and metabolic efficiency

2.2.1 $38\ \text{mol} \times 11\ \text{kcal/mol} = 418\ \text{kcal}$ ($38 \times 46 = 1748\ \text{kJ}$).
2.2.2 $418/686 \times 100$ (or $1748/2870 \times 100$) $= 61\%$.

Cold drinks, hot drinks

2.3.1 $25 \times 0.60/60 = 0.25\ °\text{C}$.
2.3.2 $25 \times 0.6 = 15\ \text{kcal}$.
2.3.3 $(15\ \text{kcal})/(100\ \text{kcal/h}) = 0.15\ \text{h}$

Oxygen and glucose in blood

2.4.1 $200/22.4 = 9\ \text{mmol/l}$
2.4.2 $90 \times 10/180 = 5\ \text{mmol/l}$.
2.4.3 $2 \times 45/100 = 0.9\ \text{mmol/(l·h)}$.
2.4.4 (a) $0.9\ \text{mmol/(l·h)} \times 5\ \text{l} \times 24\ \text{h/day} = 108\ \text{mmol/day}$,
(b) $108/1000 \times 180 = 19\ \text{g/day}$.

Basal metabolic rate

2.5.1 $1700/4.8 = 354$ l/day ($7000/20 = 350$ l/day).
$(354 \text{ or } 350)/(24 \times 60) =$ nearly 0.25 l/min.

2.5.2 (a) (1700 kcal/day)/(3.8 kcal/g) = 447 g of glucose/day.
Alternatively, (7000 kJ/day)/(16 kJ/g) = 438 g of glucose/day.
(b) (1700 kcal/day)/(9.3 kcal/g) = 183 g of fat/day. Alternatively, (7000 kJ/day)/(39 kJ/g) = 179 g of fat/day.

2.5.3 $8640 \times 1000/86\,400 = 100$ W.

Oxygen in a small dark cell

2.6.1 (1600 l)/(16 l/h) = 100 h.

2.6.2 Half of the original percentage of oxygen in the air = 10% or 10.5%.

2.6.3 10% of 760 = 76 mmHg.

Energy costs of walking, and of being a student

2.7.1 (a) $240 - 90 = 150$ kcal ($1000 - 377 = 623$ kJ),
(b) $150/3.8 = 39$ g. ($623/16 = 39$ g).

2.7.2 $9.8/30 \times 100 = 33\%$.

2.7.3 $450/150 = 3$ h.

Energy requirements for many activities and walks of life are given by Durnin & Passmore (1967).

Basal metabolic rate in relation to body size

2.8.1 The man: $1700/70 = 24$ kcal/(kg·day) ($7000/70 = 100$ kJ/(kg·day).
The mouse: $4.8/0.03 = 160$ kcal/(kg·day) ($20/0.03 = 667$ kJ/(kg·day).

2.8.2 If the small mammal has mass M, surface area S and specific metabolic rate m, then $M \times m \propto (37 - 17) \times S$. The corresponding equation for the large mammal, with temperature x, is $1000M \times m \propto (x - 17) \times 100S$. Therefore $x = 17 + (37 - 17) \times 1000/100 = 217\,°\text{C}$. Compare this calculation with that of Kleiber (1961) mentioned in the Preface.

2.8.3 $70 \times 24.2 = 1694$ kcal/day ($293 \times 24.2 = 7091$ kJ/day).

2.8.4 $70 \times 5.6 = 392$ kcal/day ($293 \times 5.6 = 1641$ kJ/day).

2.8.5 $5623 \times 70/100\,000 = 3.9$ kcal/(kg·day).
($5623 \times 293/100\,000 = 16.5$ kJ/(kg·day)).

Because of the logarithmic relationship in equation (2.3) and the huge range of mammalian body sizes, it is useful to plot specific BMR and body mass using logarithmic axes. The data for human, mouse and whale (2.8.1 and 2.8.5) are shown in Fig. 1.

For a good source on scaling and body size, see Schmidt-Nielsen (1984).

Drug dosage and body size

2.9.1 $1700A/4.8 = 354A$ ($7000A/20 = 350A$).

2.9.2 $2333A/354A$ (or $2333A/350A$) = nearly 7.

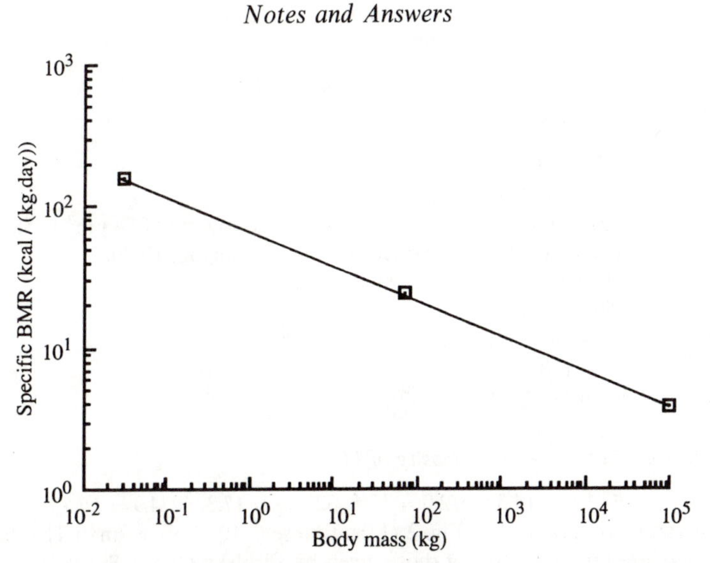

Fig. 1. Relationship between specific basal metabolic rate (specific BMR) and body mass: data for human and mouse (2.8.1) and for whale (2.8.5). Note the logarithmic scales.

For the sad story of the elephant, and comments on the dosage, see West, Pierce & Thomas (1962) and Harwood (1963).

Further aspects of allometry – life span and the heart

2.10.1 Number of beats = $241M^{-0.25} \times 6 \times 10^6 M^{+0.20} = 1.4 \times 10^9 M^{-0.05}$ beats.
2.10.2 $1^{-0.05} = 1.0$.
2.10.3 $241 \times 0.35 = 84$ beats/min.
2.10.4 $11.8 \times 2.3 = 27$ years.
Body size, physiological time and longevity are discussed by Lindstedt & Calder (1981).

The contribution of sodium transport to metabolic rate

The 'recent estimate' that 20% of the resting metabolic rate is used for sodium transport is given by Clausen, Van Hardeveld & Everts (1991).
2.11.1 20% of 24 (100) = 4.8 kcal/(kg·day) (20 kJ/(kg·day)).
2.11.2 4.8/3.9 × 100 = 123% (20/16.5 × 100 = 121%).

Production of metabolic water in human and mouse

2.12.1 16 mol/day × 18 g/mol = 288 g/day = 288 ml/day.
2.12.2 400 ml/day × 7 = 2800 ml/day.

3. Cardiovascular system

Erythrocytes and haematocrit (packed cell volume)

3.1.1 $2/5 \times 100 = 40\%$.
3.1.2 $0.91 \times 41.5 = 37.8\%$.
3.1.3 Box volume = $60 \times 2.4 = 144\ \mu m^3$. A volume of $84\ \mu m^3$ is 58% of this. (Calculation adapted from A. C. Burton, 1965).
3.1.4 $44/(5.4 \times 10^6) \times 10^9/100 = 81.5\ \mu m^3$.
3.1.5 $156/84 = 1.86$.
3.1.6 $(84 - 30) \times 0.9/(156 - 30) = 0.39\%$.
3.1.7 $84 + 180 = 264\ \mu m^3$.
3.1.8 $84/264 = 0.32$.

Optimum haematocrit – the viscosity of blood

3.2.1 $20/1.4 = 14.3$, $35/2.0 = 17.5$, $52/3.0 = 17.3$, $62/4.9 = 15.5$.
These values are graphed in Fig. 2 (1 centipoise = 10^{-3} N·sec/m^2.) The data used here are representative of those given by Pirofsky (1953). Stone, Thompson & Schmidt-Nielsen (1968) show curves of this kind for five mammalian species and point out that the camel has both the lowest normal haematocrit (27%) and also a 'hump' corresponding to the lowest optimum haematocrit (about 30%). It is gratifying when a calculation shows some feature of the human or animal body is optimal, but not every feature is. Calculations like those of 3.2.1 were first applied to the blood of land snails (R. F. Burton, 1965), where the respiratory pigment is not contained in corpuscles; the concentration of pigment is distinctly suboptimal.

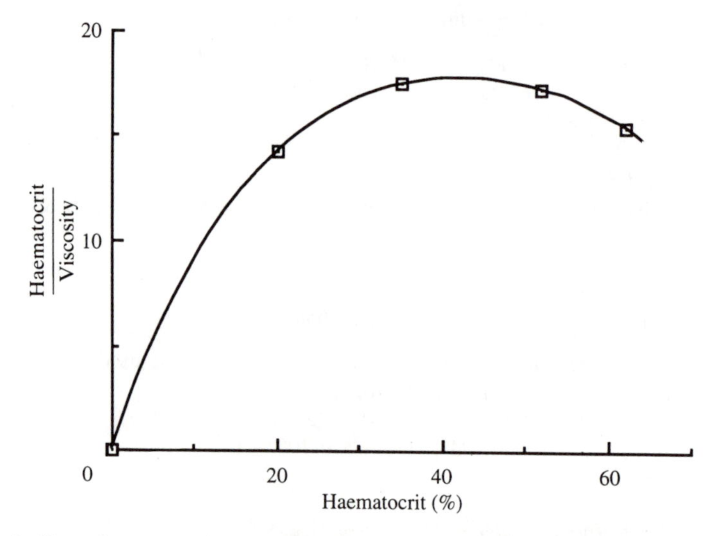

Fig. 2. Data from question 3.2.1, showing that the ratio of haematocrit (percentage) to viscosity (centipoise) is maximal (and optimal) at a near-normal value of the haematocrit.

Peripheral resistance

3.3.1 $1/3$ – or $1/3 \times 108/100 = 0.36$ if allowance is made for the change in pressure.

3.3.2 As the two ventricles have the same output, only the pressures need to be compared: $(12 - 5)/100 \times 100 = 7\%$.

Blood flow and gas exchange

3.4.1 $(5\ \text{l})/(5\ \text{l/min}) = 1$ min.

3.4.2 (a) $90\ (\text{cm}^3/\text{sec})/4.5\ (\text{cm}^2) = 20$ cm/s,
(b) 0.02 cm/s.

3.4.3 $(0.05\ \text{cm})/(0.02\ \text{cm/sec}) = 2.5$ s.

3.4.4 $A - V = (0.25\ \text{l/min})/(5\ \text{l/min}) = 0.05\ \text{l/l} = 50\ \text{ml/l}$.
$V = 200 - 40 = 150$ ml/l.

3.4.5 $200 - 2.8/20 \times 1000 = 60$ ml/l.

Arteriolar smooth muscle – the Law of Laplace

3.5.1 $(735\ \text{mmHg}\cdot\text{cm}^2/\text{kg}) \times (5\ \mu\text{m})/(15\ \mu\text{m}) \times (3\ \text{kg/cm}^2) = 735$ mmHg.

3.5.2 $(735\ \text{mmHg})/(150\ \text{mmHg}) = 4.9$.

Extending William Harvey's argument

3.6.1 $0.05\ (\text{l/min}) \times 20\ (\text{min}) = 1$ l.

The work of the heart

This topic is commonly treated graphically and in terms of single beats; the present treatment is chosen for the simplicity of the calculations.

3.7.1 $5 \times 100 \times 0.032 = 16$ cal/min. $5 \times 100 \times 0.13 = 65$ J/min.

3.7.2 $(18\ \text{cal/min})/(130\ \text{cal/min}) \times 100 = 14\%$. Alternatively, $(73\ \text{J/min})/(534\ \text{J/min}) \times 100 = 14\%$.

3.7.3 $0.5 \times 1000 \times 0.2^2/133 = 0.15$ mmHg.

3.7.4 $(0.6^2/3 + 2/3 \times 0)/1 = 0.12\ \text{m}^2/\text{sec}^2$.

4. Respiration

When not to correct gas volumes for temperature, pressure, humidity and respiratory exchange ratio

4.1.1 $(750 - 47) \times 283/\{(750 - 9) \times 310\} = 0.866$.

4.1.2 $100(1 - 0.866) = 13.4\%$.

4.1.3 3.2%.

4.1.4 $(79 + 15.8 + 3.95) \times 100/100 = 98.75\%$.

4.1.5 $3.95/(21 - 15.8) = 0.76$.

Dissolved O_2 and CO_2 in blood plasma

4.2.1 (0.03 mmol/l·mmHg) × (40 mmHg) = 1.2 mmol/l.
4.2.2 0.0014 × 100 = 0.14 mmol/l.

P_{CO_2} inside cells

4.3.1 56 × 1/20 = 2.8 mmHg.
4.3.2 43 + 2.8 = 45.8 mmHg.
4.3.3 43 + (56 − 10) × 0.8/20 = 44.8 mmHg.

Alveolar gas tensions at sea level and at high altitude

4.4.1 100/(150 − 40) = 0.9 l.
4.4.2 (350 × 40 + 150 × 0)/500 = 28 mmHg.
4.4.3 333 × 21/100 = 70 mmHg.
4.4.4 (70 − 40/0.9) = 25.6 mmHg.

Why are alveolar and arterial P_{CO_2} close to 40 mmHg?

4.5.1 Warm water: (160 − 0)/25 = 6.4 mmHg.
Cool water: (160 − 0)/35 = 4.6 mmHg.
(Rahn, 1966; Burton, 1973)
4.5.2 0.8 × (150 − 0) = 120 mmHg.
4.5.3 (a) 1.0 (i.e. all the metabolic rate!)
(b) 0.1 (10% of the metabolic rate: Burton, 1973).

Water loss in expired air

4.6.1 (a) 15 000 × 47/760 × 0.8 = 742 g,
(b) 15 000 × 37.7/760 × 0.8 = 595 g.
4.6.2 742 − 595 = 147 g.
4.6.3 15 000 × (37.7 − 15)/760 × 0.8 = 358 g.

Renewal of alveolar gas

4.7.1 1 − 0.12 = 0.88.
4.7.2 $(0.88)^2 = 0.77$.
4.7.3 Log $G = -\log 2$, so $G = 0.5$, in accordance with the definitions of G and $t_{1/2}$.
It is an important feature of any exponential time course that the rate of change (dy/dt) of a variable (y) is proportional to the instantaneous value of that variable (i.e. $dy/dt = \pm ky$). For an exponential decrease, the negative sign applies.

Variations in lung dimensions during breathing

4.8.1 3/2.5 = 1.2.
4.8.2 $\sqrt[3]{1.2} = 1.063$.
4.8.3 $1.063^2 = 1.13$.

See Gil *et al.* (1979) for the changes in the microscopical appearance with lung volume.

4.8.4 $(1 - 1/1.10) \times 100 = 9\%$.

Lung structure – branching of the airways

4.9.1 $0.25^3/2 = 0.008\ mm^3$, or $4\pi/3 \times 0.25^3 = 0.0082\ mm^3$.

4.9.2 $27 \times 1\,000000/0.0082 = 330\,000000$.

Published estimates of the number of alveoli have ranged from 7×10^7 to 7×10^8, and further. A useful discussion of airway branching is that of Thurlbeck & Wang (1974).

4.9.3 28.

4.9.4 0.5.

4.9.5 $2^4 = 16$.

With estimates of the total number of alveoli and their average dimensions, one may estimate the total aveolar surface area and the reward is some aptly impressive figure. However, because of the surface irregularities, there is much the same problem in defining the area as there is in defining the length of a coastline (Mandelbrot, 1983).

Surface tensions in the lungs

Note that Laplace's formula given here differs from that given in Section 3.5; for a cylinder, $P = T/r$ and for a sphere, $P = 2T/r$.

4.10.1 $15 \times (70\ mN/m)/(100\ \mu m) = 10.5\ mmHg$.

4.10.2 $15 \times 25/100 = 3.75\ mmHg$.

For a rebuttal of the entrenched idea that the surface tension in the alveoli can be zero, see Bangham (1991).

Pulmonary lymph formation and oedema

4.11.1 $25 - 7 = 18\ mmHg$. The flow would be into the capillaries.

4.11.2 $25 - 7 - 9 = 9\ mmHg$.

4.11.3 $-9\ mmHg$.

4.11.4 $9 \times 120/15 = 72\ mN/m$.

There is more to the roles of surfactant and of surface irregularities in the stabilizing of alveolar fluid. Guyton, Moffatt & Thomas (1984) give a good account of this interesting topic.

The pleural space

4.12.1 From capillaries to pleural space:
$24 - 25 + 6 + 4 = 9\ mmHg$.

4.12.2 From pleural space to capillaries:
$-7 + 25 - 6 - 4 = 8\ mmHg$.

Pleural liquid pressures have been reviewed by Agostoni & D'Angelo (1991).

5. Renal function

The composition of the glomerular filtrate

5.1.1 $300 \times 19.3 = 5790$ mmHg.
5.1.2 $60/5700 \times 100 = 1\%$.
5.1.3 $(180\,\text{l}) \times (0.01$ to $0.1\,\text{g/l}) = 1.8\text{–}18$ g.

The influence of colloid osmotic pressure on glomerular filtration rate

5.2.1 $25/(1 - 0.2) = 31.25$ mmHg.
5.2.2 New net filtration pressure $= (45 - 10 - 26) = 9$ mmHg.
$(9 - 7)/7 \times 100 = 28.6\%$.

Glomerular filtration rate, inulin and PAH clearance, drug clearance

It is usual to begin the topic of renal clearance with a general formula:

$$\text{renal clearance} = \frac{U \times V}{P},$$

where V is the rate of urine flow and U and P are concentrations of any given substance in urine and plasma respectively. This formula is less intuitively obvious than equation (5.3) and students commonly memorize the formula with a lack of understanding that is evident from their inability to suggest appropriate units for clearance values (e.g. ml/min).

5.3.1 $(0.25\,\text{mg/min}) \times 1000/(4\,\text{mg/l}) = 6.25$ ml/min, or
$(360\,\text{mg/day})/(4\,\text{mg/l}) = 90$ l/day.
5.3.2 (a) $0.693 \times 14\,000/125 = 77.6$ min,
(b) 2×77.6 min $= 155.2$ min.
5.3.3 $2 \times 77.6 = 155.2$ min.
5.3.4 $0.693 \times 14\,000/660 = 14.7$ min.

The concentrating of tubular fluid by reabsorption of water

5.4.1 $1/(1 - 2/3) = 3$.
5.4.2 $125/1.25 = 100$.
5.4.3 100×5 mmol/l $= 500$ mmol/l.

Urea – clearance and reabsorption

5.5.1 $450/4.5 = 100$ l/day.
5.5.2 $(1 - 70/125) = 0.44$.
5.5.3 2/3.
5.5.4 $(450\,\text{mmol})/(45\,\text{l}) = 10$ mmol/l.

Sodium and bicarbonate – rates of filtration and reabsorption

5.6.1 $180 \times 150/1000 = 27$ mol.
5.6.2 $27 \times 58.5/1000 = 1.58$ kg.

5.6.3 $180 \times 25 = 4500$ mmol/day.
5.6.4 $90/4500 \times 100 = 2\%$.

Is fluid reabsorption in the proximal convoluted tubule really isosmotic?

5.7.1 1–10 mosmol/kg.
5.7.2 No.
For the data on rat nephrons, and for further discussion, see Schafer (1982).

Work performed by the kidneys in sodium reabsorption

5.8.1 $27/29 =$ nearly 1 mol.
5.8.2 $1 \times 22.4 = 22.4$ l/day.
5.8.3 $22.4/350 \times 100 = 6.4\%$.
5.8.4 Threefold greater.
The relationship between sodium reabsorption and oxygen consumption was studied by Thaysen, Lassen & Munck (1961). There is a distinction between the total work done by the kidneys in elaborating the urine (and reflected in oxygen consumption) and the thermodynamic work needed to produce all the differences in solute concentrations between plasma and urine. The latter is only about 1% of the total. If the kidneys were to produce a urine identical in composition to plasma, then the thermodynamic work would be zero, but the work of reabsorption would still be substantial.
5.8.5 $15 \times 1000/200 = 75$ ml/min.
The mitochondrial content of rat kidneys was estimated by Pfaller & Rittinger (1980). For measurements of the maximal oxygen consumption of mitochondria from mammalian skeletal muscle, see Schwerzmann *et al.* (1989).
5.8.6 $(15 \text{ ml/min})/(300 \text{ g}) = 0.05$ ml/min per g.
5.8.7 Previous answer $\times 2.6 = 0.13$ ml/min per g.
5.8.8 $0.13 \times 100/18 = 0.72$ ml oxygen/min per g.
5.8.9 $23 + 40/5 = 31$ l/day.

Mechanisms of renal sodium reabsorption

5.9.1 $29/5.6 = 5.2$.
5.9.2 $1/2 \times 100/6 = 8\%$ or $1/3 \times 100/6 = 5.6\%$.

Autoregulation of glomerular filtration rate, glomerulotubular balance

5.10.1 (a) $126 - 124 = 2$ ml/min,
(b) $125 - 123 = 2$ ml/min.

Renal regulation of extracellular fluid volume and blood pressure

5.11.1 $125 - 1 = 124$ ml/min.
5.11.2 $(135 - 131)/1 = 4$.

Daily output of solute in urine

In Table 5.1 the amounts of cation, in terms of equivalents, exceed the amounts of anion, but in any urine sample they must be equal (Section 6.5). This discrepancy arises partly because the data are from various sources.

5.12.1 528 to 1278 mosmol/day.

5.12.2 (8 to 17) × 6.25 = 50 to 106 g/day.

The flow and concentration of urine

5.13.1 (750 mosmol/day)/(300 mosmol/l) = 2.5 l/day.

5.13.2 2 × 300 mosmol/l.

5.13.3 (a) 750/15 = 50 mosmol/l,
(b) 750/0.6 = 1250 mosmol/l.

The data of questions 5.13.1 to 5.13.3 lie on the curve of Fig. 3 which corresponds to a constant solute load of 750 mosmol/day.

5.13.4 2 − 0.5 = 1.5 l/day.

5.13.5 1200/300 − 1200/1200 = 3 l/day.

The previous two calculations are from Harvey (1974).

5.13.6 1200/1000 = 1.2 l/day.

5.13.7 The (1.2 − 0.5) = 0.7 l of urine in a day is more than the 600 ml of sea water.

5.13.8 0.5 − 600/9000 = 0.43 l/day.

5.13.9 (a) (30 mosmol/l) × (25 l/day) = 750 mosmol/day,
(b) 50 × 25 = 1250 mosmol/day.

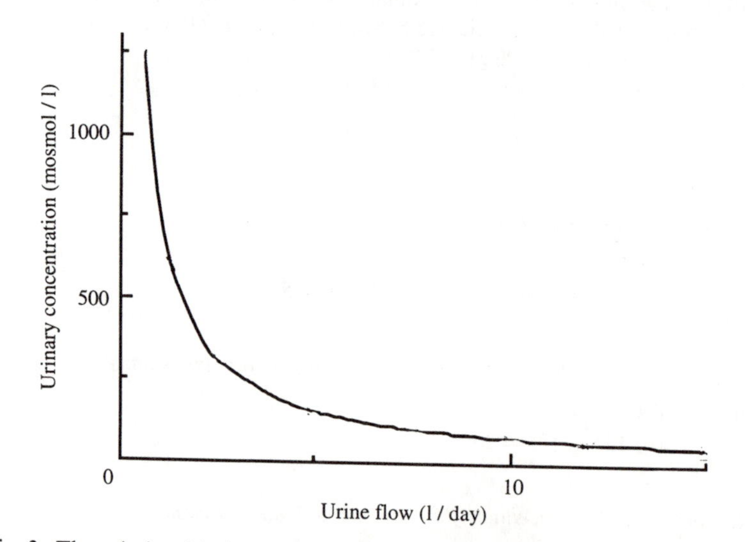

Fig. 3. The relationship between urinary concentration and flow rate for a constant solute output of 750 mosmol/day.

Beer drinker's hyponatraemia

5.14.1 $240/6 = 40$ mosmol/l.

For an account of beer drinker's hyponatraemia see Hilden & Svendsen (1975). Danish beer contains 1–2 mmol/l of sodium.

The medullary countercurrent mechanism – applying the principle of mass balance

5.15.1 $359.4 \times 1000/1199.5 = 299.6$ mosmol/l.
5.15.2 $30 + 4 - 20 - 0.5 = 13.5$ ml/min.
5.15.3 (a) $300 + (200 \times 20 - 900 \times 0.5)/30 = 418$ mosmol/l,
(b) $300 + 3550/200 = 318$ mosmol/l.
5.15.4 $12\,540/16.5 = 760$ mosmol/l.

6. Body fluids

A note on osmoles and osmotic pressure

The osmotic pressure of an ideal solution depends on the total concentration of solutes in mol/kg water. Thus, 1 mol of solute in 22.4 kg of water at 0 °C ideally exerts an osmotic pressure of 1 atmosphere, or 760 mmHg (just as 1 mol of ideal gas occupies 22.4 l when under a pressure of 1 atmosphere at 0 °C). In the case of a salt such as NaCl, the dissociated sodium and chloride contribute separately to the total.

Physiological solutions are too concentrated to show ideal behaviour and the interactions of the various solutes reduce their total osmotic effectiveness. Thus, the concentration of an ideal solution that has the same osmotic pressure as a real one containing just 150 mmol NaCl/kg water is about 280 rather than 300 mmol/kg water. One way of dealing with this discrepancy is to use another unit, the osmole, such that the 'osmolality' of that solution is 280 mosmol/kg water.

Re-wording an earlier statement to define this unit, 1 osmol of solute in 22.4 kg of water at 0 °C exerts an osmotic pressure of 1 atmosphere, or 760 mmHg. The number of osmoles of a solute may be calculated from the number of moles by multiplying the latter by an empirical factor called the 'osmotic coefficient'. This varies with such things as the nature of the solute and the concentration of the solution; for NaCl in the above solution the osmotic coefficient is $280/300 = 0.93$.

The osmotic pressure also increases in proportion to the absolute temperature. Thus, at 37 °C the same 1 osmol of solute in 22.4 kg of water has an osmotic pressure of $760 \times (273 + 37)/273 = 863$ mmHg. What to remember is that 1 mosmol/kg water at body temperature exerts an osmotic pressure of 19.3 mmHg.

Osmoles are generally only used in the context of total solute concentration, and especially where that relates to osmotic pressure. However, osmolalities are generally calculated from the colligative properties of depression of freezing point or depression of vapour pressure.

Another drink of water

6.1.1 (a) $[1 - 49.5/(49.5 + 0.5)] \times 100 = 1\%$,
(b) 1% *of* 300 = 3 mosmol/kg water.

Cells as 'buffers' of extracellular potassium

6.2.1 (30 mmol/l)/(15 l) = 2 mmol/l. 4.5 + 2 = 6.5 mmol/l.
6.2.2 (a) $150 \times 30 = 4500$ mmol,
(b) $5 \times 15 = 75$ mmol.
6.2.3 $30/4500 \times 100 = 0.7\%$ – almost impossible to demonstrate.

Assessing movements of sodium between body compartments – practical difficulty

6.3.1 (150 + 3) mmol/(1 + 0.027) kg = 149 mmol/kg water.

The role of bone mineral in the regulation of extracellular calcium and phosphate

For more on the solubility relations of octocalcium phosphate, see Driessens, Verbeeck & van Dijk (1989).
6.4.1 9/7 = 1.29, 8.5/4.5 = 1.89, 9/4.5 = 2, 8/6 = 1.33. The range is thus 1.29–2.0.
6.4.2 The concentration of free calcium would rise by 0.45 mmol/l. If it started at 1.3 mmol/l, the percentage rise would be 35%, much as for phosphate.
6.4.3 The concentration of calcium would rise by $(1.5 \times 0.1 \times 30/100) = 0.045$ mmol/l, i.e. 0.45% of 10 mmol/l.
6.4.4 $10 - 1.5 \times 0.1 = 9.85$ mmol/

The principle of electroneutrality

6.5.1 $$\frac{10^{-6}\text{F/cm}^2 \times 0.07\text{ V}}{1 \times 10^5 \text{ coulombs/mol}} = 7 \times 10^{-13} \text{ mol/cm}^2$$
(since volts × farads = coulombs).
6.5.2 (a) $14 \times 10^{-6}/10 = 14 \times 10^{-7}$ mmol/l,
(b) $14 \times 10^{-6}/0.5 = 28 \times 10^{-6}$ mmol/l.
6.5.3 $\{144 + 4 + (2 \times 1) + (2 \times 0.5)\} - (102 + 28 + 1 + 18) = 2$ mequiv/l.
In relation to the accurate analysis of plasma, remember that the concentrations of bicarbonate and chloride change with carbon dioxide tension, along with the net charge of the plasma proteins, through buffering and through chloride/bicarbonate exchange across erythrocyte membranes. All three therefore differ as between arterial and venous plasma, and in blood samples exposed to air. The total concentration stays constant, however, except inasmuch as there are small shifts of water between erythrocytes and plasma.
6.5.4 $(138 + 3 + 3 + 6) - (70 + 5) = 75$ mequiv/l.
6.5.5 $18 + 135 + 1 - 78 - 16 = 60$ mmol/kg water.
There is clearly a large quantity of anions not accounted for. These include protein (mainly haemoglobin) and phosphates (including 2,3-diphosphoglycerate). The calculation cannot reveal whether or not there are other cations present.

Donnan equilibrium

6.6.1 $141.3 + 18 = 159.3$ mmol/kg water.
6.6.2 $150/159.3$ or $141.3/150 = 0.942$.
6.6.3 $61.5 \times -0.026 = -1.6$ mV.

Colloid osmotic pressure

6.7.1 $300 \times 19.3 = 5790$ mmHg.
6.7.2 Solution (2), by 0.54 mmol/l.
6.7.3 $(1 + 0.54) = 1.54$ mosmol/kg water.
6.7.4 $1.54 \times 19 = 29$ mmHg.
6.7.5 $1/1000 \times 68\,000 = 68$ g/kg water.
For a discussion of colloid osmotic pressures and protein net charge, see Burton (1988).

Molar and molal concentrations

6.8.1 $0.99(1 - 0.75 \times 70/1000) = 0.94$.
6.8.2 $141/0.94 = 150$ mmol/kg water – significantly different from 141.
6.8.3 $0.99(1 - 0.75 \times 360/1000) = 0.72$.
6.8.4 $150 \times 0.72 = 108$ mmol/l – a much greater discrepancy than for plasma.

Osmolarity and osmolality

6.9.1 281.5 mmol/l.
6.9.2 $0.93/0.94 = 1.0$.

Gradients of sodium across cell membranes

6.10.1 (10–13 kcal/equiv)/(23.1 kcal/volt·equiv) $\times 1000/3 = 144$–188 mV.
6.10.2 Somewhat above $(90 + 40) = 130$ mV.
6.10.3 $61.5 - (-90) = 151.5$ mV.
6.10.4 $61.5(1 + 0.30) - (-90) = 170$ mV.
6.10.5 $61.5 - 2/3 \times (-94) - 1/3 \times (-90) = 154$ mV.

Membrane potentials – simplifying the Goldman equation

6.11.1 (a) $0.05 \times 15/150 \times 100 = 0.5\%$,
(b) 5%, likewise.
6.11.2 (a) $61.5 \log(4.5/150) = -93.7$ mV,
(b) $61.5 \log[(4.5 + 1.5)/150] = -86.0$ mV,
(c) $61.5 \log[(4.5 + 10.5)/150] = -61.5$ mV.

7. Acid–base balance

pH and hydrogen ion activity

7.1.1 4×10^{-8} mol/l or 40 nmol/l. (Since $\log 4 = 2 \log 2 = 0.6$, $10^{0.6} = 4$. Alternatively, antilog $0.6 = 4$.)

7.1.2 1 μm^3 contains $10^{-7} \times 6 \times 10^{23} \times 10^{-15} = 60$ hydrogen ions.
7.1.3 $3 \times 0.1^2 \times 4 = 0.12\ \mu m^3$.
7.1.4 $60 \times 0.12 = 7$.
There is another problem in defining pH as $-\log[H]$. If $[H] = 10^{-7}$ mol/l, what happens to the units when you take the logarithm? Logarithms can only be taken of dimensionless numbers, so that the units of concentration (mol/l) need to be removed by dividing by unit concentration (1 mol/l).

The CO_2–HCO_3 equilibrium: the Henderson–Hasselbalch equation

The prime in pK'_1 is to indicate that this is an 'apparent' equilibrium constant – i.e. it features the concentrations of bicarbonate and carbon dioxide ($S \cdot P_{CO_2}$) rather than their activities.
7.2.1 (a) $pH = 6.1 + \log(12/1.2) = 7.1$,
(b) $pH = 7.1 + 0.3 = 7.4$.
7.2.2 It falls by 0.3 unit.
7.2.3 It falls by about 0.3 unit, the point being that 83/39 is close to 2.
7.2.4 $6.1 - (-1.52) = 7.62$.
7.2.5 $7.4 - \log(24/40) = 7.62$. This answer has to be the same as the previous.
7.2.6 $(1.023 - 1) \times 100 = 2.3\%$.
7.2.7 $0.05 \times 0.2 = 0.01$.
Therefore the answer is again 2.3%. For a discussion of pK'_1 in the Henderson–Hasselbalch equation, see Burton (1987).

There are other ways of graphing acid–base data. The graph of bicarbonate concentration against P_{CO_2} has the advantages of utilizing as axes the two determinants of pH and yielding for plasma *in vitro* a curve that relates to the carbon dioxide dissociation curve.

Intracellular pH and bicarbonate

7.3.1 $61.5 \times (7.0 - 7.4) = -24.6$ mV.
7.3.2 $-24.6 - (-70) = +45.4$ mV.
7.3.3 $61.5 \times \{pH_i - pH_e\} = -70$.
$pH_i - pH_e = -1.14$.
$pH_i = 7.4 - 1.14 = 6.26$.
7.3.4 $26/2 = 13$ mmol/kg water.
7.3.5 1.0; the two quantities are equal.

Why bicarbonate concentration does not *vary with P_{CO_2} in simple solutions lacking non-bicarbonate buffer*

7.4.1 $(24.00000 + 0.00010 - 0.00004) - (0.00005 - 0.00008) = 24.00009$ mmol/l.
Not measurably different.
7.4.2 $[HCO_3]$ is virtually unchanged, but [H] is doubled.
Therefore, the $P_{CO_2} = 2 \times 40 = 80$ mmHg.

Carbonate ions in body fluids

7.5.1 (a) Antilog $(7.8 - 9.8) = 0.01$,
(b) 0.1,
(c) 1.0.
7.5.2 $7.62 + \log(20/1.3) = 7.62 + \log 15.4 = 8.81$.

Buffering of lactic acid

7.6.1 (a) Antilog $(4.6 - 6.6) = 0.01$
(b) 0.001, similarly
7.6.2 $7.4 - 0.1 = 7.3$
7.6.3 Previous answer minus $\log 5 = (7.3 - 0.7) = 6.6$.

The role of intracellular buffers in the regulation of extracellular pH

7.7.1 The pH rises by $\log(28/25) = 0.05$.
7.7.2 $7 \times 3/15 = 1.4$ mmol/l.
That the amount of bicarbonate leaving the erythrocytes is influenced by the final concentration in the extracellular fluid, and therefore by other sources of bicarbonate, may be understood by analogy. If a warm object (≡erythrocytes) is dropped into water (≡extracellular fluid) that is cooler, the extent to which the object loses heat (≡bicarbonate) is greater if the volume of water is greater. The object loses less heat if other warm objects (≡nucleated cells that also release bicarbonate) are dropped in with it.
7.7.3 A factor of 2 also.
7.7.4 None.
For a more detailed discussion of the movements of bicarbonate between cells and extracellular fluid in disturbances of acid–base balance, see Burton (1992).

The role of bone mineral in acid–base balance

7.8.1 $25 + 1 = 26$ mmol/l.
7.8.2 A rise in pH of $\log(26/25) = 0.017$ unit.

The postprandial alkaline tide

The figures for the rates of gastric acid secretion are from Bowman & Rand (1980); pharmacology books are the readiest sources of such data, because of their relevance to antacids. Does the size of a typical antacid tablet (sodium bicarbonate, perhaps) seem right for those rates?
7.9.1 $25 + 28/14 = 27$ mmol/l.
7.9.2 $\log(27/25) = \log 1.08 = 0.03$ pH unit.
7.9.3 $(40–110) \times 1.5/24 = 2.5–6.9$ mequiv.

8. Nerve and muscle

For more on 'number numbness' see Hofstadter (1985) who taught that physics class.

Myelinated axons – saltatory conduction

8.1.1 (1.5 mm)/(100 mm/ms) = 0.015 ms.
Some of the textbook figures referred to depict action potentials in squid axons or mammalian nerve fibres at low temperature, which are therefore slower, but this does not explain the difference.
8.1.2 At least 1.5 m.
8.1.3 (1 m)/(10 m/s) × 1000 − 10 = 90 ms.

Non-myelinated fibres

8.2.1 $120^2/4.8 = 3000$ μm (3 mm).
Whether the myelinated fibres in the central nervous system with diameters of 0.2–1 μm conduct more or less rapidly than non-myelinated fibres of the same diameter is not known.
8.2.2 $2.2 \times \sqrt{100} = 22$ m/s.

Musical interlude – a feel for time

8.3.1 1000/11 = 91 ms.
8.3.2 The reaction to a sound would probably take between (150 − 50) = 100 ms and (250 − 30) = 220 ms, both times being longer than the previous answer.

Muscular work – chinning the bar, saltatory bushbabies

8.4.1 9.8 × 0.4 = 3.9 J/kg body mass.
8.4.2 3.9/70 × 100 = 5.6%.
The muscles, kindly dissected and weighed by Dr A. Chappell, were the latissimus dorsi, teres major, pectoralis major (less the clavicular head), biceps, coracobrachialis, brachialis, and brachioradialis.
8.4.3 5.6/2 = 2.8%.
The other route to the conclusion that short people should generally find it easier to lift themselves than do tall ones is as follows. The ease with which one holds oneself up, rather than lifts oneself, increases with the areas of cross-section of the relevant muscles and decreases with body mass. For a body of given shape, areas are proportional to the square of any given linear dimension L, i.e. to L^2, while body mass is proportional to L^3. The ease with which one supports oneself is therefore proportional to L^{-1}. As to the lift, two other factors are relevant, the height of the lift and, opposing this, the lengths of the relevant muscles. Both are proportional to L. Ease of lifting is therefore proportional to $L^{-1} \times L/L = L^{-1}$.
8.4.4 2.1 × 9.8 = 20.6 J/kg.
8.4.5 20.6/70 × 100 = 29.4%.
The calculations on the bushbaby derive from a paper on muscle work by Alexander (1992) and references therein.

Phosphocreatine in muscular contraction

8.5.1 $70/50 = 1.4$ mmol/kg.

Calcium ions and protein filaments in skeletal muscle

8.6.1 (a) $\pi \times 0.5^2 = 0.79\ \mu m^2$,
(b) $1 \times 0.79 = 0.79\ \mu m^3$.
8.6.2 $0.8 \times 10^{-15} \times 10^{-7} \times 6 \times 10^{23} = 48$.
8.6.3 $1/(0.9 \times 10^{-3}) = 1100$.
8.6.4 $0.8 \times 1100 = 880$.
8.6.5 $880/48 = 18$.
8.6.6 $47\,520 \times 4/10\,000 = 19$.
8.6.7 $3 \times (450$ nm$)/(45$ nm$) = 30$.

Appendix A: a useful logarithm

9.1 $3.00/10 = 0.300$ (slightly low).
9.2 $3 \times 0.301 = 0.903$.
9.3 $1 - 0.301 = 0.699$.
9.4 $(2 - 0.301)/2 = 0.85$. (log 7 is actually 0.845)

Index